The Camera Obscura

Frontispiece. Images of the sun through openings in foliage, from *The Forces of Nature* by A Guillemin (1872). This illustration of camera obscura images created in nature is a Victorian engraver's attempt to illustrate the observation made by Aristotle.

The Camera Obscura

A Chronicle

John H Hammond

Adam Hilger Ltd, Bristol

British Library Cataloguing in Publication Data

Hammond, John H.
The camera obscura.
1. Camera obscura
I. Title
681′.4 QC787.C/

ISBN 0-85274-451-X

Published by Adam Hilger Ltd, Techno House, Redcliffe Way, Bristol BS1 6NX.
The Adam Hilger book-publishing imprint is owned by The Institute of Physics.

Text set in 12/14pt Plantin by Quadraset Ltd, Radstock, and printed in Great Britain by J W Arrowsmith Ltd, Bristol.

For Mother

and

GMB, ECS, DGHH

Lo, here a camera obscura is presented to thy view . . . if thou art perfectly at leisure for such trivial amusements, walk in and view the wonders of my inchanted garden.

Erasmus Darwin

Preface

The camera obscura is now hardly known though at one time it was invaluable to astronomers, a must for travellers and a dilettante's toy. Historians of photography dismiss it at 1839 yet it continued as an instrument in its own right being manufactured until the beginning of the century. It was used by the Royal Air Force in the last war and today is still to be found as a drawing office aid and acting as a tourist attraction.

The camera obscura is a very simple optical instrument which has changed little since the eighteenth century when it was part of the instrument maker's stock in trade. In 1914 it was made for the services who used it for reconnaissance but since that time camera obscuras have been constructed only for individual requirements. This chronicle of its long history is told through the writings of those who made, used or merely described the instrument. They are mainly isolated excerpts and do not lend themselves to a narrative style but are ordered in centuries.

However, a separate chapter is devoted to a review of its use by artists, a particularly interesting aspect of the history of the camera obscura.

For the sake of clarity and ease of reading the term 'camera obscura' (plural camera obscuras) is used throughout though it was not adopted until the seventeenth century. In past centuries writers often abbreviated the term to 'camera', an abbreviation which would now lead to confusion.

John H Hammond

Acknowledgments

The material for this book was derived from many sources, not the least being books for which I am indebted to authors and librarians, especially the staff at Slough Public Library, the Science Library at Kensington and the Royal Photographic Society. I must thank museum curators, local government officers, business managers, editors and a great number of people who gave time to answer my enquiries. To friends and colleagues who responded to my quest I am grateful. I wish to thank Mr Dall of Luton and Mrs Partridge at the Castle Museum, Nottingham, for showing me their camera obscuras. My thanks are due also to Dr D B Thomas, Dr A Sahiar and Mr John Ward and their colleagues at the Science Museum, and to Mr Graeme Fyffe at the Science Library.

It has been a happy experience to share the enthusiasm with which Neville Goodman of Adam Hilger Ltd has converted my typescript bundle to this pleasing volume.

Most of all am I grateful to Miss Elaine Spratt and Dr David Halstead who so patiently commented on the final draft.

Thanks are due to the following for permission to reproduce illustrations:

Her Majesty the Queen: 34, 37
American Heritage Publishing Co. Inc: 88
American Photographic Book Publishing Co. Inc., New York: 59
Avon County Library, Bristol Reference Library: 73
Barr and Stroud Ltd: 70
British Library: 10, 23, 36, 48
Christie's South Kensington Ltd: 102
Country Life: 95

Mr Horace Dall: 97
Dover Publications: 7
Dumfries Museum and Mr David Hope: 68, 69
East Sussex County Library: 74
Mr A D Gracey: 98
Halco-Sunbury Ltd: 100
Harvard University Collection of Historical Scientific Instruments: 54
Mr Lionel Hughes: 66
India Office Library: 39
Kilmarnock Standard: 67
Department of Medical Illustration, Royal Infirmary, Manchester: 41
Kent County Library: 90, 91
Mauritshuis Gallery, The Hague: 33
The Trustees of the Museum of Applied Arts and Sciences, Sydney, Australia: 82
Museum of the History of Science, Oxford: 49
National Gallery of Art, Washington, DC (Gift of Mrs Robert W Schuette): 38
Rijksmuseum, Amsterdam: 46
Royal Australian Air Force: 99
Royal Society, London: 47
Science Museum, London (Crown Copyright): 5, 12, 13, 14, 16, 17, 18, 19, 20, 21, 22, 24, 27, 28, 29, 30, 31, 35, 44, 51, 52, 53, 55, 56, 57, 58, 60, 61, 65, 75, 76, 78, 79, 80, 81, 83, 85, 86, 87
Swansea Museum: 89
Thames and Hudson Ltd, London: 48
Victoria and Albert Museum, London: 6, 43, 64
Vintage Cameras Ltd: 26, 77
Walker Art Gallery, Liverpool: 11, 40

Contents

The appendixes at the end of each chapter contain references too brief to be included in the text, and information acquired during the preparation of the book.

Early Centuries

The Locked Treasure Room

Chinese texts of the fifth century BC indicate quite clearly that the writers had already discovered by experiment that light travels in straight lines. The philosophers Mo Ti and Chuang Chou commented on shadows in a manner which might seem naïve to us, but what they said were and still are fundamental truths, the basic facts of science. For instance, it had already been observed that a shadow does not move by itself; it only moves if either the object or the light source is moved.[1] It had also been noted that if there are two light sources, there will be two corresponding shadows. The Chinese had made observations on the sizes of shadows; when a rod is held in the beam of a fixed light source to create a shadow on a screen, the length of the shadow will change from long to short as the rod is titled from the vertical to slant towards or away from the screen.

Mo Ti was the first to record the creation of an inverted image with a pinhole in a screen, which has been variously translated as a 'collecting place' (the rays of light meeting at the pinhole), etc, but perhaps most appropriately as a 'locked treasure room'. He was aware that objects reflect light ('shining forth') in all directions, and that rays from the top of an object, when passing through a hole, will produce the lower part of an image. Mo Ti also noted that only those light rays passing through the pinhole will make an image; other rays will strike the screen above or below the hole.

Although these observations show an understanding of image formation, no further mention was made of the camera obscura effect until early in the ninth century AD, when Tuan Chheng Shih referred to an image of a pagoda.

He thought that it was inverted because it was near to the sea, which seemed to produce that effect. It is thought that the inverted pinhole image could have been confused with reflections of pagodas seen on the surface of a lake, where images always appear upside down to an observer. However, Shen Kua later corrected this false explanation by referring to the crossing of rays of light at a pinhole, which he compared to an oar in a rowlock: when the oar handle is down, the blade is up.

In the tenth century Yu Chao-Lung used model pagodas to make images on a screen with a small hole to observe the direction and divergence of rays of light. It is curious that in spite of these experiments and observations, no geometric theory on image formation evolved from the conclusions. Shen Kua made an analogy between the inverted image and the nature of man: he suggested that there were some people who, like the inverted image, will so misunderstand a

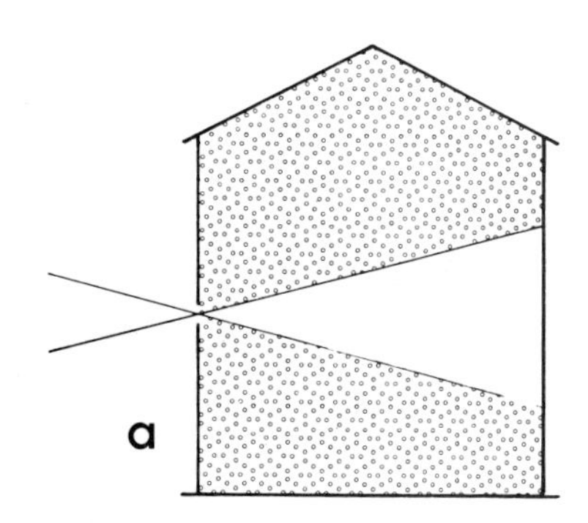

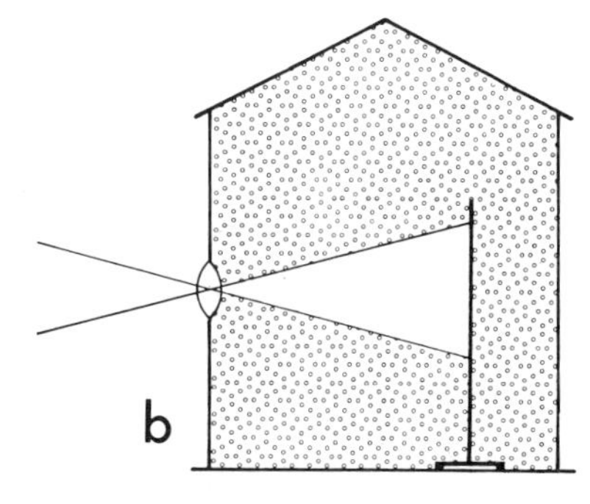

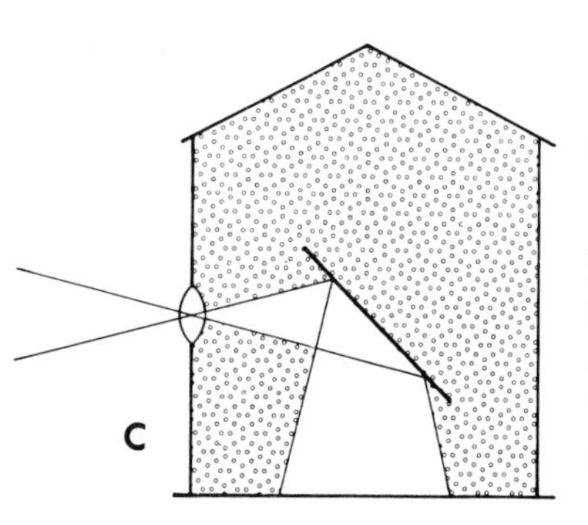

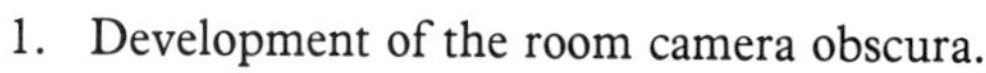

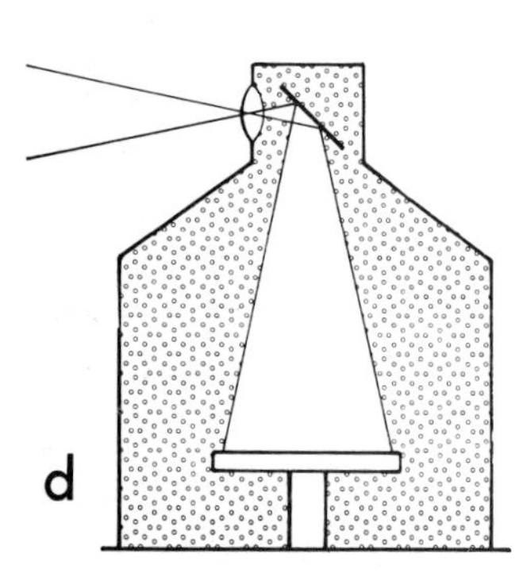

1. Development of the room camera obscura.
(a) A small hole in the shutter of a darkened room will make an inverted image of an outdoor scene. The image has no exact position of focus.
(b) A lens makes a much brighter inverted image but a movable screen is required in order to obtain a sharp picture.
(c) When reflected by a large mirror, the image appears upright to an observer. Sharp focus is achieved by moving the mirror backwards and forwards.
(d) A lens and small mirror in a turret on the roof of a room projects an image onto a table where it may be viewed upright. Sharp focus is obtained by sliding the lens in its mount or by moving the table up and down. Some camera obscuras, when erected on a hill dispense with the need for focusing because there are no near objects in view.

situation as to think right is wrong. Indeed, there are some, he said, who will be so fixed in their ideas that they could hardly avoid seeing things upside down.

It would be interesting to learn exactly how the Chinese first discovered the image-making properties of a small hole. It is likely that the image of a pagoda could have been first noticed in the room of a house or hut, perhaps when the room was shuttered against the heat of midday. A small gap would allow the rays from a sunlit scene with a pagoda to pass through and make an image on the opposite wall.

It was Aristotle who, during the fourth century BC, first saw the totally natural occurrence of the camera obscura effect when on the ground below a tree he noticed many crescent-shaped images of the sun during an eclipse, and discovered that they were caused by the small spaces between the leaves.[2] He then extended his observations to the holes of a sieve, to the small holes made by crossing the fingers of

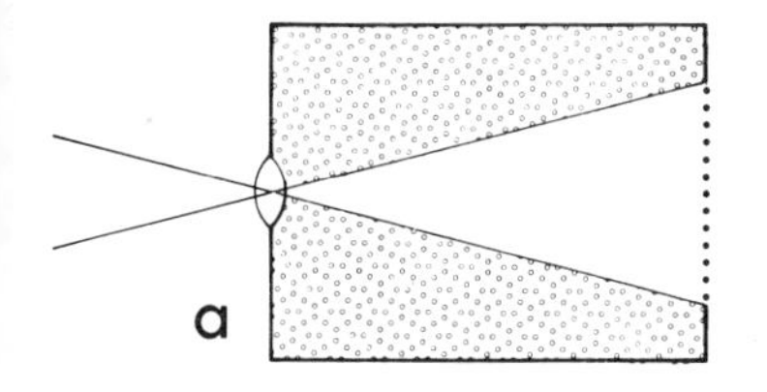

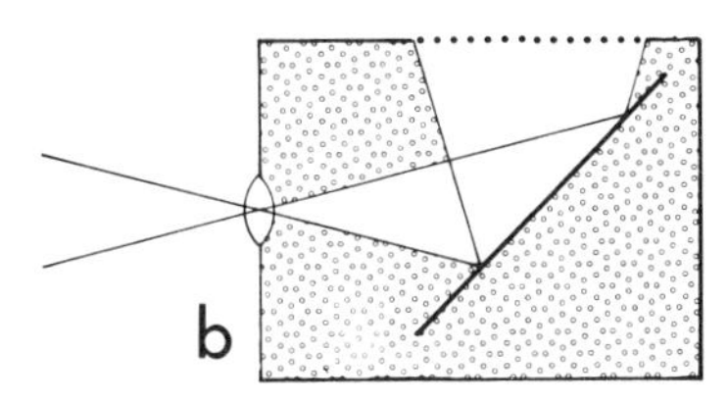

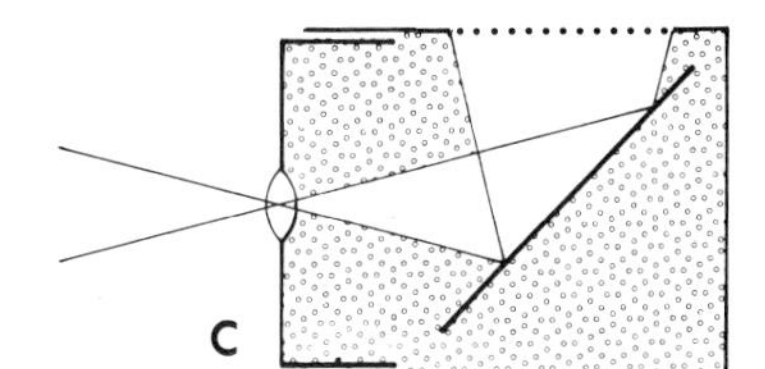

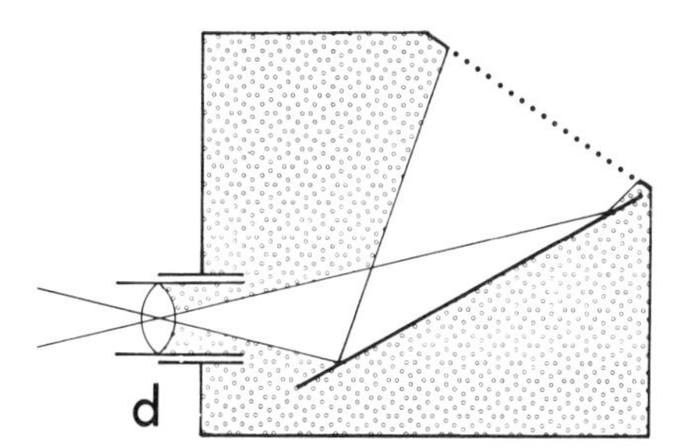

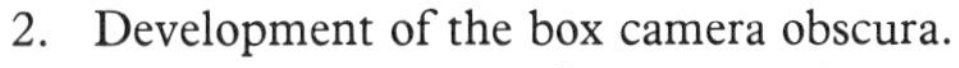

2. Development of the box camera obscura.
(a) A lens in one side of a box produces an inverted image on the opposite side which is made of a translucent material such as either ground glass or thin paper.
(b) A mirror at 45° in the box reflects the image, giving an upright picture on the ground glass at the top of the camera obscura.
(c) Similar to (b) but the lens is fitted to a box sliding in the main body of the camera obscura in order to obtain sharp focus of near objects.
(d) A redrawing of one of Zahn's seventeenth-century camera obscura (p38) in which the ground-glass screen is at an angle facing the observer. The lens is mounted in a sliding tube for focusing.
(e) Diagram of William Storer's eighteenth-century 'Royal Accurate Delineator' (p78). The large-aperture compound lens is mounted in a sliding box for focusing. Another (plano-convex) lens fixed below the ground glass increases the image brightness, especially at the corners of the screen.

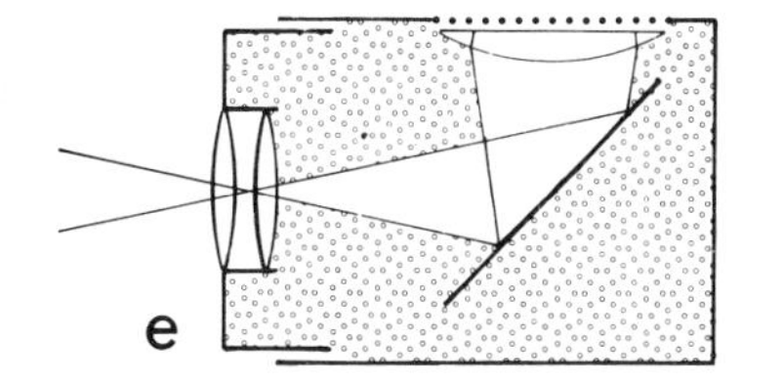

one hand over the other, and to the small gaps in wicker work. Aristotle was not unduly perturbed by the images, nor the fact that they were upside down; he simply explained them as being the result of a cone of light from the sun with its apex at the hole, which in turn made a cone of light on the other side, so creating in this case an image of the sun. What

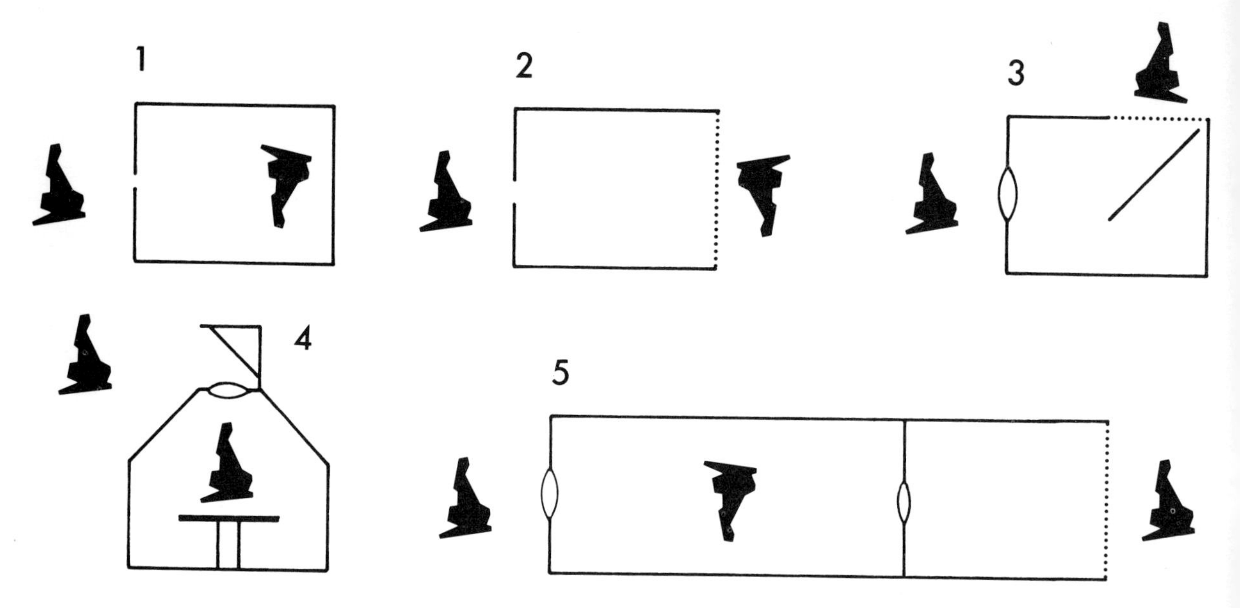

3. Orientation of the image in the camera obscura.
(1) A room-type camera obscura with a small hole or a lens presents an image which is upside down and reversed.
(2) A simple box camera obscura with a small hole or a lens presents an upside-down image on a ground-glass screen. A tracing of the image on paper is correct when turned through 180°.
(3) A reflex camera obscura presents an image on a ground-glass screen which is upright but completely reversed. A tracing of the image is correct when the paper is turned over or when the drawing is viewed in a mirror.
(4) A room camera obscura with a lens and mirror in a turret, and tent, book-form or pyramid camera obscuras present images which are correctly orientated.
(5) A simple box camera obscura with a second image-forming lens presents a correct image, as in (4). This construction was suggested by Kircher and later by Leyser, but it is extremely clumsy and the image is dim.

did concern him, however, was that regardless of the shape of the hole—whether it was produced by the fingers of one hand over the other or a hole in a wicker basket—the image of the sun was always round or crescent-shaped during an eclipse. Even when the hole was quite irregular, as between leaves, the shape of the image was always the same. He could find no satisfactory explanation to this observation, and it remained unresolved until the sixteenth century. Over the years many philosophers attempted to solve what became known as one of Aristotle's problems.

The Arabian physicist and mathematician of the tenth century, Alhazen, also made experiments on image formation.[3,4] He arranged three candles in a row, and between them and a wall he set up a screen with a small hole. He noticed that the candle to the right of the hole made an image to the left on the wall, and that the image of the left-hand candle appeared to the right; from this Alhazen deduced the linearity of light. He noted that the images were formed only by small holes; large holes simply produced a patch of light whose outline was the same shape as the hole.[5] The experiment with the candles required some form of darkened room, and although Alhazen referred to the reversal of the image from left to right, he made no comment on the fact that it was inverted.[3] This appears curious in view of the previous ninth-century reference to images of pagodas being upside down.

These early references to Chinese and Greek observations on shadows and images are so close in time that one may wonder whether their discoveries were actually made independently. There is an inclination to consider ancient Eastern and Western civilisations as having been separate, but there may have been some early scientific communication, particularly since a flourishing silk trade had been established between Rome and China by the first century AD.[6]

During the compilation of this history no references to the camera obscura effect in the eleventh and twelfth centuries were found.

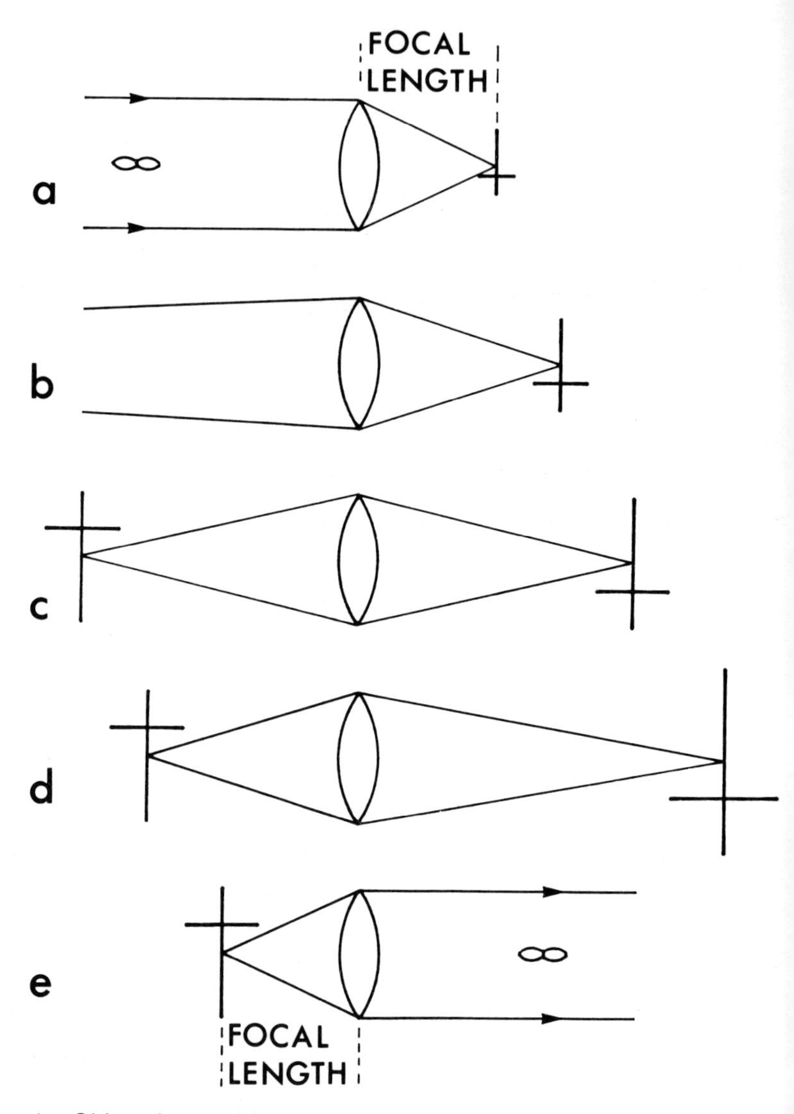

4. Object, lens and image.
(a) A lens makes an image at its focal length of objects a long distance away; about 1000 yards or more—so-called infinity, or optical infinity.
(b) When the distance between the object and the lens is decreased the image is made further away than the focal length.
(c) When the object is two focal lengths away from the lens the image is also two focal lengths away, and is the same size as the object.
(d) If the object is closer than two focal lengths from the lens the image is made at a greater distance and is larger than the object.
(e) When the object is placed at the focal length of the lens the image is a long distance away, at infinity.

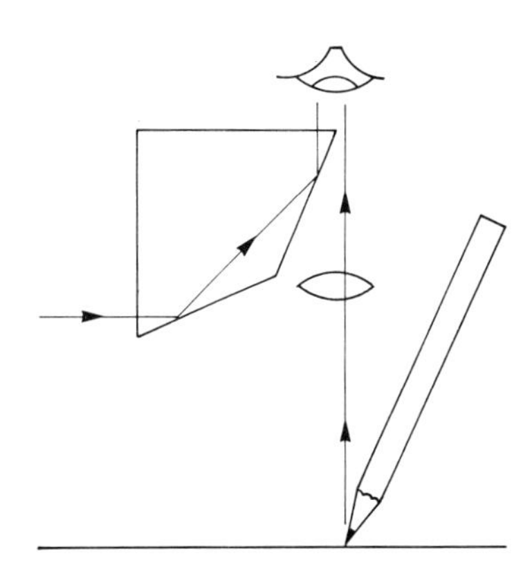

5. The camera lucida, a drawing aid which received considerable attention from optical designers during the nineteenth century. It was not an easy instrument to use; the eye of the draughtsman had to be placed so that the centre of the pupil was directly above the edge of the prism. A reflection of the object or view was seen in the prism by half of the eye and the pencil point by the other half. As the images fused on the retina it was possible to trace the outline of the reflection. If the object was some distance away, a lens was introduced to enable the eye to focus the distant reflection and the near pencil point at the same time.

References

1. Needham J 1954 *Science and Civilisation in China* vol. 4 (London: Cambridge University Press)
2. Hett W S 1936 *Aristotle—Problems* (London)
3. Ronchi V 1970 *The Nature of Light* (London: Heinemann)
4. Bradbury S and Turner G L'E 1967 *Historical Aspects of Microscopy* (Cambridge: Heffer)
5. Eder J M 1945 *History of Photography* (New York: Columbia University Press)
6. Grant M 1960 *The World of Rome* (London: Weidenfeld and Nicholson)

The Thirteenth and Fourteenth Centuries

Lenses, spectacles and concave mirrors had been produced before the beginning of the thirteenth century, but a lens had not replaced the small hole or pinhole in a camera obscura.

At this time astronomers were making observations of the solstices using a gnomon to cast a shadow along a scale. In an attempt to obtain greater accuracy they later used a longer, thinner gnomon, but the shadow produced then became very diffuse and difficult to determine, so this was of no advantage. The Chinese astronomers Kuo Shuo-Ching and Yang Huan then discovered that a pinhole could be used to create a clearer image of the tip of the gnomon along the scale, and by this means they were able to obtain extremely accurate measurements of solstice shadows.[1]

During the thirteenth century, however, interest in scientific matters declined rapidly in China, and was not revived until Jesuit missionaries were sent to the East in the sixteenth century. In the West, on the other hand, there was a great deal of enquiry and activity in the natural sciences.

In Britain, Roger Bacon, one of the more prolific workers of the period, wrote on the reflection and refraction of light and the use of mirrors and magnifying glasses, thus extending Alhazen's work on image formation. Bacon described how mirrors could be used to produce an aerial image of an object or scene in such a way that on approaching the image, an observer would see nothing.[2] In order to see people walking below a window, he suggested the use of an inclined mirror in front of an aperture of a camera obscura. Although Bacon has often been given credit for the invention of the camera obscura, it seems that he did not actually describe one;[3] it is perhaps for this reason that Goethe suggested that many of the experiments described by Bacon resulted from an inventive mind and were never actually carried out.[4] Dr Freind, however, writing in the early eighteenth century, reported that Bacon had spent some £2000 in twenty years on optical and mathematical instruments, 'a prodigious sum for such a sort of expences in those times.' It is possible that the camera obscura was so well known among scientists at the time that Bacon could have taken it for granted and assumed there was no need for a formal description.

Roger Bacon, Guillaume de Saint-Cloud, Witelo (Witeck or Vitellio, a Polish physicist) and John Peckham (or Pecham, who was an Archbishop of Canterbury and is thought to have been a pupil of Bacon) all mentioned the camera obscura as a means of observing eclipses of the sun.[5] Another contemporary, Villeneuve, was some sort of miracle-worker or magician who used one to entertain his friends. Witelo based his book *Optics* largely on Alhazen's work, but gave a clearer explanation of image formation. Saint-Cloud, in an almanac of 1290, wrote of many people who had been dazzled or partially blinded for some days after staring too intently at an eclipse of the sun, and suggested making a hole 'about the size of that in a barrel for drawing off wine' in the roof of a house facing an eclipse, with a flat board or screen opposite the hole about twenty feet away. He explained that the image would be larger but weaker the further away the screen was placed from the hole, and also that the image would be a reversal of the actual happening. Witelo and Peckham both attempted to solve Aristotle's problem, but without success.

The magic and showmanship of Villeneuve was not unlike a shadow play or a cinema; he often arranged for players to enact some exciting war-like or murderous episode in an area of bright sunshine in view of a room to be used as a camera obscura. The room was darkened and had a small hole in the wall or shutter facing the actors, and during the play he arranged for a group of people outside the room to make appropriate noises, such as the din and clash of swords, or screams and blasts from trumpets. His guests may have been delighted with the entertainment but it is likely that some of them, in an age of superstition, would have taken fright.

During the fourteenth century Levi Ben Gershon, also known as Leon de Bagnois, a Jewish philosopher and mathematician from Arles in Provence, also observed eclipses by means of a camera obscura and remarked that no harm had been done to his eyes.[6] Stained glass was available as a filter to protect the eyes, but very few astronomers used it because the often unpolished surfaces, as well as particles and bubbles in the glass, degraded the image.[7]

References

1. Needham J 1954 *Science and Civilisation in China* vol. 3 (London: Cambridge University Press)
2. Waterhouse J 1901 *Photographic Journal*
3. Eder J M 1945 *History of Photography* (New York: Columbia University Press)
4. Freind, Dr John 1726 *History of Physik* (London)
5. Potoniee G 1936 *History of the Discovery of Photography* (New York: Tennant and Ward)
6. Gernsheim H 1969 *History of Photography* (London: Oxford University Press)
7. King H C 1955 *History of the Telescope* (London)

The Fifteenth Century

Leo Battista Alberti, an Italian painter and architect, is sometimes associated with the camera obscura, especially in dictionaries and books on art, but from other accounts it appears that he neither owned nor even used one. He made a 'show box' which, perhaps through indifferent translation, has occasionally become confused with the camera obscura.[1-3]

A similar show box was made by Gainsborough, and this is now exhibited at the Victoria and Albert Museum in London. This was a wooden box with a peep-hole or lens at one side, with the opposite side open, over which was placed a painting made in thin transparent colour on glass or paper.

6. Gainsborough's show box, now at the Victoria and Albert Museum in London. Crown Copyright, Victoria and Albert Museum.

This was viewed through the peep-hole with candles behind the painting providing the illumination. The illusion of perspective was enhanced by the single-eye view, the isolation of the picture and the transmitted light which added brilliance to the painting.

7. Two drawing aids illustrated by Dürer (1471–1528). The upper engraving shows a framed grid and a sighting rod which provided the artist with a single-eye viewpoint. The drawing paper was ruled with a grid which enabled him to transfer what he saw through the frame onto the paper. The lower engraving shows a sighting tube at the end of a cord attached to the wall which also gave the artist a single-eye viewpoint and an effectively longer viewing distance. This reduced distortion due to foreshortening in the drawing.

The frequent difficulty of translation has resulted in several different explanations of Alberti's box.[4] It may have been a perspective box (an example of which can be seen at the National Gallery in London), or a box containing a mirror in which the viewer could see a reflection of a painting, or it may have been a box with a sheet of glass between the peep-hole and the object. The artist painted the object on the sheet of glass, as illustrated by Dürer.

Alberti also made a device to gain accuracy in drawing comprising a frame with equally spaced horizontal and vertical strings forming a grid, through which a scene or object could be viewed. By having drawing paper ruled with a similar grid it was then relatively easy for the artist to reproduce any view seen through the frame.

Leonardo da Vinci also became involved with the problem of image formation using small holes.[5,6] His drawings of light rays quite clearly represent a camera obscura, showing cones of light from the object converging on a pinhole, and diverging on the other side to make an inverted image. He described how to make a suitable hole in a thin sheet of iron and suggested receiving the image on a translucent screen so that it could be viewed from the rear. He noted that objects reflect rays of light in all directions, and that images will be formed at any place by the passage of light rays through a small hole onto a screen. This was, perhaps, the most important observation made by Leonardo and he illustrated it by drawing a camera obscura with two pinholes, each of

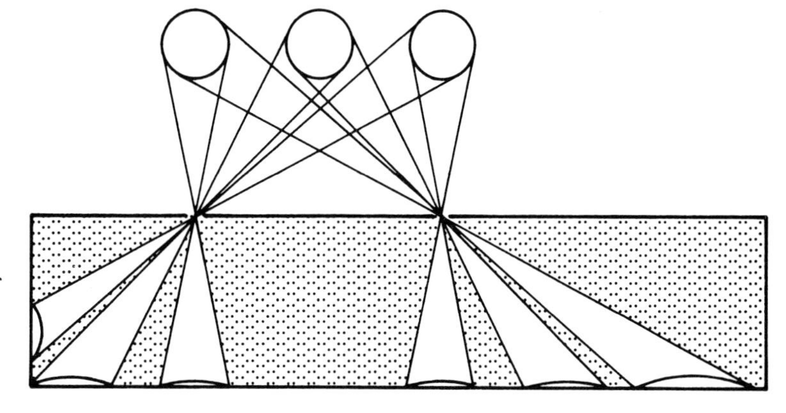

8. An illustration from the notebooks of Leonardo da Vinci (1452–1519) demonstrating the reflection of light in all directions. A camera obscura with two pinholes produces two sets of images, each pinhole making an image of the same object. (Redrawn from Richter.)

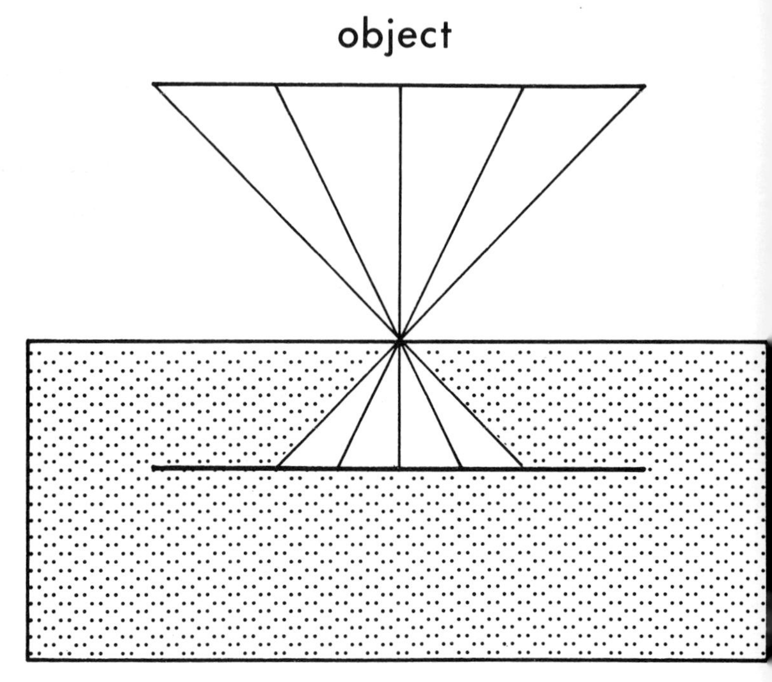

9. Leonardo da Vinci described the eye by comparing it with a camera obscura. He suggested that the rays from an object illuminated by the sun should be received on a sheet of white paper held near the hole of a camera obscura. This may be the first reference to an image viewed from behind a translucent screen. (Redrawn from Richter.)

which produced a set of images from one group of objects. It will be remembered that the reflection of light in all directions had been previously observed by the Chinese when they noted that rays also struck the surface above and below the pinhole.

References

1. Bradbury S and Turner G L'E 1967 *Historical Aspects of Microscopy* (Cambridge: Heffer)
2. *Encyclopaedia Britannica* 1910
3. Clark K 1949 *Landscape into Art* (London: Murray)
4. Ivins W M Jr 1973 *On the Rationalisation of Sight* (New York)
5. Eder J M 1945 *History of Photography* (New York: Columbia University Press)
6. Richter J P and Richter I A 1970 *The Literary Works of L. da Vinci* (compiled and edited from the original manuscripts) (London: Phaidon)

The Sixteenth Century

Considerable technical advances were made in the evolution of the camera obscura during the sixteenth century. Spectacles for improving failing eyesight had been in use for some time, and had become common among those able to read. One of the earliest references to the use of a lens in place of a small hole in the camera obscura was made by Girolamo Cardano, a professor of mathematics at Milan, who was also an astute gambler and had a somewhat chequered career both as an academic and a family man.[1-3] He had some leanings toward medicine and suggested that compulsive gambling should be considered as a disease. The translation of Cardano's numerous publications has proved difficult because of the changing nature of language. There is some doubt as to whether he was referring to a lens or a concave mirror, but he was familiar with both, and mentioned the possibility of seeing people passing by in the street by means of a concave mirror. Like Villeneuve, Cardano played the part of showman and magician using a darkened room to project outdoor scenes of actors with fearful monsters and accompanied by suitable noises.

A somewhat clearer description of the use of a biconvex lens was given by Daniele Barbaro of Venice, an architect and patriarch of Aquileia.[2-4] He referred to what was probably the only choice of spectacles at the time: concave for young people who tended to be near-sighted, and convex for the elderly. An old man's spectacle glass, he said, should be used for a camera obscura. It is to Barbaro that we owe two very important though different aspects of the camera obscura. He suggested reducing the aperture of a lens in order to make a clearer or sharper image, and pointed out that there was a best position or focus for receiving an image on a piece of paper. When this was achieved, he said, one could take a pencil and draw on the paper the outline of the image, and on the drawing one could shade in and colour as of nature. This may be the earliest reference to the use of the camera obscura for drawing.

Another Venetian, Giovanni Battista Benedetti, also referred to the use of a biconvex lens and the fact that the inverted image could be corrected by placing a mirror at an angle of 45° to reflect the image onto a table in the dark room.[3] This arrangement of lens and plane mirror to produce an upright image is used in the camera obscura today. The main difference, however, is that in the sixteenth century the camera obscura was a room of a house made dark by closing the shutters, whereas today it is a structure designed specifically for the purpose. Ignatio Danti, a Florentine mathematician and astronomer, was also concerned about the inverted image of the camera obscura; he used a concave mirror to make the image upright, but the method was clumsy and impractical.[4] The correction by means of a second biconvex lens had to wait until the seventeenth century.

The solution to Aristotle's problem of an irregular hole giving a round image of the sun was given by Francisco Maurolycus, a mathematician who may be considered a forerunner to Kepler.[5] Maurolycus applied himself to the theory of image formation both by the eye and by a small aperture without a lens. It appears that he did little or nothing with the camera obscura as an instrument, but did suggest that it could be used for solar observation to avoid harming the eyes. So keen were astronomers to observe eclipses and sunspots at this time that they often stared at the sun without any eye protection, and many reported confused vision or even temporary blindness for several days as a result. John Greaves, a professor of astronomy at Oxford, recorded that after measuring the sun's diameter he hurt his eyes, 'insomuch that for some days after, to that eye with which I observed, there appeared, as it were, a company of crows flying together in the air at a good distance.'[5] The prevalence of the intense though painful desire to watch solar happenings is obvious from the number of references to the advantage of using a camera obscura for the purpose.

Erasmus Reinhold, a German mathematician, explained how to use a camera obscura to observe eclipses and described how he had seen those that had occurred in 1544 and 1545.[3,4,6] His pupil Reinerus Gemma-Frisius, a Dutch physician and mathematician, published the earliest yet discovered illustration of a camera obscura, in which an image of the 1544 solar eclipse is shown projected onto the wall of a room. A similar arrangement was used by the astronomers Copernicus, Tycho Brahe, Moestlin, Kepler and Fabricius.

Early in the sixteenth century Johann Fabricius and his father had measured sunspots using a simple camera obscura without a lens.[7] When looking directly at the sun with a telescope they hurt their eyes; it had apparently never occurred to them to use the telescope to make a projected image. A great number of detailed observations were published by Fabricius in 1611. Freidrich Risner also referred to the use of a camera obscura for eclipse observations and commented that it could be used for copying drawings.[4] He noted that with the aid of the camera obscura drawings could be enlarged or reduced, and that it was ideal for sketching topographical views. For this purpose he gave detailed advice on the construction of a light wooden hut to house the apparatus. He was, it seems, an enthusiastic astronomer for in connection with his observations he referred to Peckham, Maurolycus, Reinhold and Reinerus.

Leonard Digges and his son Thomas, both English mathematicians, have a somewhat tenuous connection with the

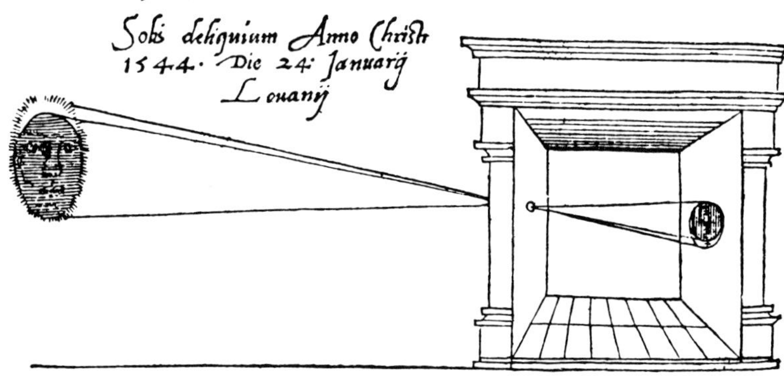

10. Reinerus Gemma-Frisius observed an eclipse of the sun at Louvain on 24 January 1544, and later he used this illustration of the event in his book *De Radio Astronomica et Geometrico* (1545). It is thought to be the first published illustration of a camera obscura.

camera obscura.[8] The father wrote a treatise on mathematics which was published posthumously by his son, who added some notes of his own. It is sometimes stated that Leonard Digges invented the camera obscura, on the basis of notes and commentaries he wrote which were similar to those of Roger Bacon.

The most flamboyant character of the sixteenth century, and perhaps of the whole history of the camera obscura, was Giovanni Battista della Porta; no one has received more acclaim for the invention of the instrument than this Neapolitan scientist.[9,10] He travelled widely throughout France and Spain, and when at home kept an open house for his fellow scientists. Although a devout Catholic, the activities and gatherings at Porta's home brought suspicion of witchcraft upon him from Rome but he managed to clear himself of any charges. Through a large circle of friends throughout Europe he was able to indulge a desire, which almost amounted to an obsession, to collect and record all sorts of scientific and natural history information, facts and lore. At the age of 23 he had accumulated sufficient material to publish a book under the title *Magiae Naturalis*, containing party tricks, domestic recipes for cleaning and cooking, and personal toiletry, to which he added some chapters on optics. This work was written in a popular, easy style and covered a wide range of subjects; it was an immediate success and became a best seller. The first edition of 1558 comprised four volumes, and the second edition (1589) had twenty; it was translated into Arabic and several European languages and there were innumerable reprintings. Porta had gathered all that was then known about the camera obscura, its construction and use, and he claimed to have made improvements which, in a good journalistic manner, he referred to as secrets. Porta described

> How in a chamber you may see hunting, battles of enemies, and other delusions,—and animals that are really so, or made by art of wood or some other matter. You must frame little children in them, as we use to bring them in when comedies are acted; and you must counterfeit stags, bores, rhinocerets, elephants, lions and what other creatures you please;

and so on, but for sheer originality he added,

> We may demonstrate the same without the light of the sun, not without wonder. Torches or lights lighted on purpose in chambers, we may see in another chamber what is done, by fitting as I have said, but the light must not strike upon the hole, for it will hinder the operation; for it is a second light (reflected) that carries the images. I will not conceal at last a thing that is full of wonder and mirth, because I am faln upon this discourse, that by night an image may seem to hang in a chamber. In a tempestuous night the image of anything may be represented hanging in the middle of the chamber, that will terrify the beholders. Fit the image before the hole, that you desire to make to seem hanging in the air in another chamber, that it is dark; let there be many torches lighted around about. In the middle of the dark chamber place a white sheet, or some solid thing, that may receive the image sent in; for the spectators that see not the sheet, will see the image hanging in the middle of the air, very clear, not without fear and terror, especially if the artificer be ingenious.'

Surely the artificer could be no more ingenious than Porta himself. Having published such a popular work it is 'not without wonder' that so many books, encyclopaedias and dictionaries have since referred to Porta as the inventor of the camera obscura.

References

1. Bradbury S and Turner G L'E 1967 *Historical Aspects of Microscopy* (Cambridge: Heffer)
2. Potonniee G 1936 *History of Photography* (New York: Tennant and Ward)
3. Waterhouse J 1901 *Photographic Journal*
4. Gernsheim H 1969 *History of Photography* (London: Oxford University Press)
5. King H C 1955 *History of the Telescope* (London)
6. Eder J M 1945 *History of Photography* (New York: Columbia University Press)
7. *Encyclopaedia Britannica* 1910, 1973
8. Digges L 1571 *Pantometria* (London)
9. *British Journal of Photography* October 1888
10. Porta G B della *Magiae Naturalis* (English trans. 1658)

The Seventeenth Century

The seventeenth century saw a considerable expansion of enquiry into the mechanism of vision, in optical theory and the advent of the telescope for astronomical observations. In this century the camera obscura was named as such for the first time and it became more widely known among English scientists.

Francis Bacon made the curious observation that the image was more apparent on a white screen than on a black one,[1] and he received a letter from Sir Henry Wotton containing a description of Kepler's tent camera obscura.[2] Many writers of books on astronomy and mathematics discussed the images produced through a small hole in a shutter of an enclosed, dark room, and other scientists confirmed the work of previous writers. For example, Bettinus in his *Apiara* said that it was Barbaro who first enlarged the hole to accommodate a lens.[3] Several of the publications during this century were in the nature of compendiums, quoting and illustrating the work of others. These were the forerunners of the many dictionaries of popular science and technology which followed in the eighteenth and nineteenth centuries.

Robert Boyle was probably the first English scientist to construct a portable camera obscura, which he demonstrated to the Royal Society.[4] Several years later he wrote an account of it for their journal, although he probably only did so out of pique at the large number of imitations. His account described a box with a lens at one end and a sheet of paper 'at a convenient distance' from it, and the image was viewed through a hole somewhere in the upper part of the box. It is not clear whether the viewing hole was in the upper side of the box or in the upper part of the end carrying the lens, although the latter would have been the better for seeing the image. There was no suggestion that the camera obscura was used for drawing, so the viewing aperture may have been merely a peephole.

Jeremiah Horrox, a poor, ill-paid curate in the village of Hoole near Liverpool, made important contributions to astronomy and was the first to observe the transit of Venus across the sun, an event which had been predicted by Kepler for the year 1639.[5] The exact date, 24 November, was calculated by Horrox who corrected many astronomical tables. The transit was due to happen on a Sunday and Horrox feared it would interrupt his clerical duties, but fortunately the event took place at 3.15 in the afternoon. In order to make the observations,

11. The founder of English astronomy. A painting by Eyre Crowe (1824–1910) of Jeremiah Horrox making the first record of the transit of the planet Venus on the afternoon of Sunday 24 November 1639.

> I described on a sheet of paper a circle whose diameter was nearly equal to 6 inches, the narrowness of the apartment not permitting me conveniently to use a larger size . . . When the time of the observation approached, I retired to my apartment, and having closed the windows against the light, I directed my telescope, previously adjusted to a focus, through the aperture towards the sun and received his rays at right angles upon the paper already mentioned. The sun's image exactly filled the circle, and I watched carefully and unceasingly for any dark body that might enter upon the disc of light.

In the Walker Art Gallery in Liverpool there is a typically Victorian reconstruction portrait by Eyre Crowe of Horrox with his telescope and the projected image of the sun.

Robert Hooke, for some time an assistant to Robert Boyle, was a scientist, lecturer and investigator of many trades and crafts.[6–8] In 1681, in order to demonstrate human vision to the Royal Society, he constructed a large conical camera

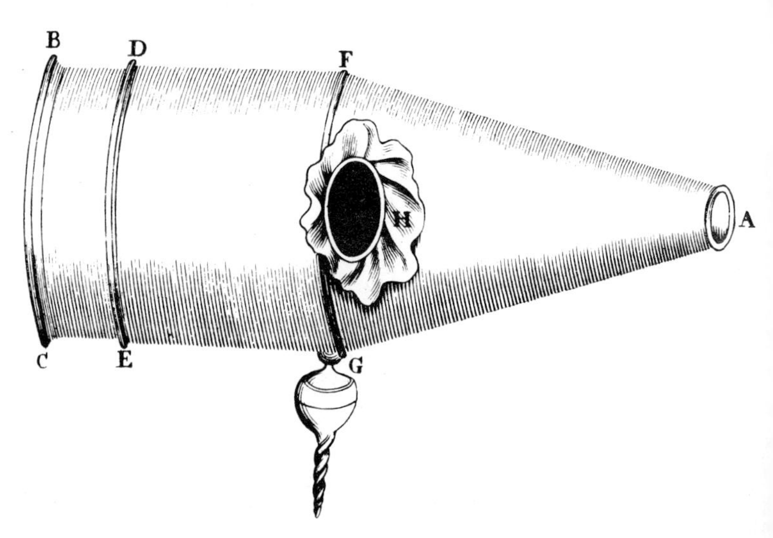

12. Robert Hooke's camera obscura which he used to demonstrate the function of the eye in his lecture to the Royal Society in 1680. The lens at a A was fitted with diaphragms to show the expansion and contraction of the pupil, while the peephole at H allowed observation of the image on a concave screen at BC. Focusing was achieved by sliding the tube containing the screen into the body DFGE. The ball and socket support at G may have been the artist's addition, since it seems inadequate for the apparatus, which was about four feet long.

obscura with a lens at the apex and a hole part-way along the cone. The base of the cone extended as a parallel tube in which another sliding tube was fitted. By this means it was possible to focus the image, but instead of doing this by moving the lens, Hooke moved the screen. The image could be seen by peeping through the hole in the side of the conical portion. Hooke said that the base should be white and concave; this was probably the first reference to the use of a concave screen to accept an image from a simple lens with its inevitable curved field of sharp focus. Some years later, in 1694, Hooke described another camera obscura which, he said, could be used to make a true drawing of anything and to which could be added shadows and light. He recommended his device as a means of improving book illustrations, particularly those of travel books which at that time were very poorly illustrated. On 21 February 1666, Samuel Pepys went to Gresham College to hear Robert Hooke lecture on the craft and trade of felt-making,[9] 'And anon alone with me [talked] about [the] art of drawing pictures by Prince Robert's rule and machine, and another of Dr. Wren's; but he says nothing doth like squares, or, which is the best in the world, like a darke room—which pleased me mightily.'

Cornelis Drebbel was an extraordinary Dutchman who spent the latter half of his life in England; he appears to have been partly an engineer, part alchemist and wholly a magician.[10] He constructed a submarine propelled by oarsmen which made the journey from Westminster to Greenwich, and he used a camera obscura as a showpiece to entertain his friends. One of these, Constantyn Huygens, spent a good deal of time with Drebbel when he was in London and often mentioned him in his letters. Huygens described the camera obscura images with considerable excitement, referring to

13. A camera obscura by Robert Hooke which he described in a paper to the Royal Society in 1694. The representation is stylised, although the general shape is similar to his demonstration model. The background scenery is a reminder of Hooke's suggestion that a camera obscura should be used to make illustrations for travel books.

them as 'the apotheosis of the optical image', and compared them with paintings made by some Dutch artists. He even implied that the painter Jan Symonszoon van der Bek, known as Torrentius, may have used a camera obscura for his pictures. Huygens' letters to his parents caused them some anxiety for they were perturbed at their son's association with the sorcerer Drebbel. His son, Christian Huygens, the physicist, astronomer and mathematician, worked on problems concerning binocular vision and used a camera obscura as a model when discussing the eye.

René Descartes wrote about the eye and vision in his book *La Dioptrique* (1637), which included an illustration showing an eye as the lens and the retina as a screen of a camera obscura.[11] A man is shown looking from behind the retina at the inverted image.

Johann Hevelius, a German astronomer with an interest in microscopy, discussed image formation and the principle of the camera obscura in his work *Selenographia* published in 1647.[1] Another astronomer, Pierre Gassendi, used a camera obscura to observe the passage of Mercury across the sun in 1631.[12]

It is to the German astronomer Johann Kepler, however, that we owe the name 'camera obscura'.[1] Writers had previously used such terms as conclave obscurum, cubiculum tenebricosum and camera clausa. Kepler learned of the camera obscura from Porta's book, and then made a thorough examination of the theory of image formation himself, extending the work of Maurolycus. Kepler often used a camera obscura in his studies and is reported to have used one in his observations of the transit of Mercury in 1607.[13] He described a method for measuring the diameters of the sun and the moon, and how to make enlarged images with the camera obscura. Kepler also demonstrated how to correct the inverted image by using a second convex lens, and how the focal distance of a lens could be reduced by

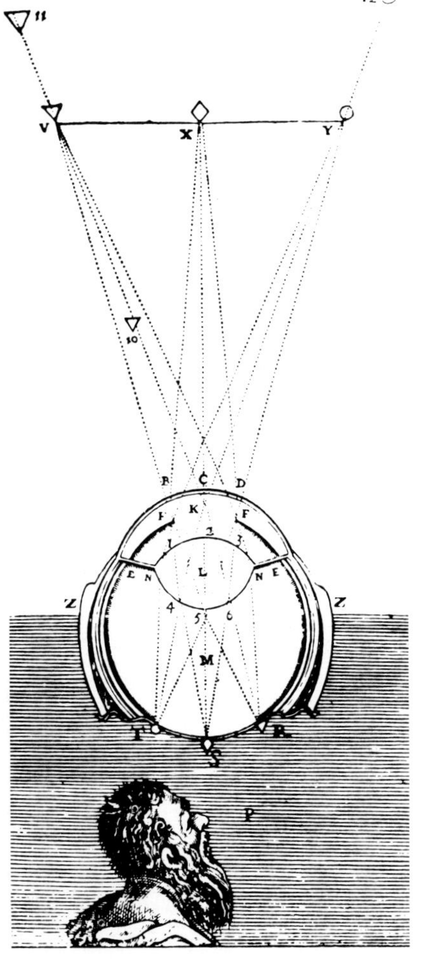

14. René Descartes' illustration of the eye as a camera obscura, 1637. The addition of an observer was merely a convention similar to that used in the illustration from Kircher (p26).

15. Kepler's suggestion of a second lens in order to produce an erect image in a camera obscura is shown in this engraving from Scheiner's *Oculus* (1619). A crude copy of this illustration, 30×35 mm in size, was printed upside down in Leurechon's *Mathematical Recreations* (1633).

interposing a negative concave lens.[14] This may be the first description of what we now know as a telephoto lens, although Thomas Digges stated that his father had produced a combination lens in as early as 1571.[8] Digges made the interesting claim that his father had put lenses together in such a way that coins scattered on the Downs could be read from afar. It was a rather loosely described arrangement of lenses, but unfortunately we have no further evidence and no apparatus to support the claim.

The English diplomat and traveller Sir Henry Wotton visited Kepler who showed him a camera obscura constructed as a tent which could be easily carried from place to place; it is now thought to have been the first of its kind.[2] It was probably the same camera obscura that Kepler had used to make a survey of Upper Austria when he occupied the post of Imperial Mathematician.[15] Wotton wrote a letter to Francis Bacon giving a full account of Kepler's tent and made the comment: 'This I have described to your Lordship, because I think there might be good use made of it for Chorography.' At that time the word was applied to topographical drawings of land for transfer and legal purposes.

Another German, Athanasius Kircher, a professor of philosophy, mathematics and oriental languages at a Jesuit college in Rome, also described a portable camera obscura.[8] His illustration is curious, although typical of the period, depicting a man inside a box which, to the same scale, would have been twelve or more feet cubed. There was a trap door in the base for him to enter and the whole rested on two horizontal poles projecting beyond the box. Presumably these were handles for two men to lift and carry the camera obscura like a sedan chair. It is more than likely that the device was no larger than a sedan chair, but it may even have been smaller so that only the head and arms were inserted.

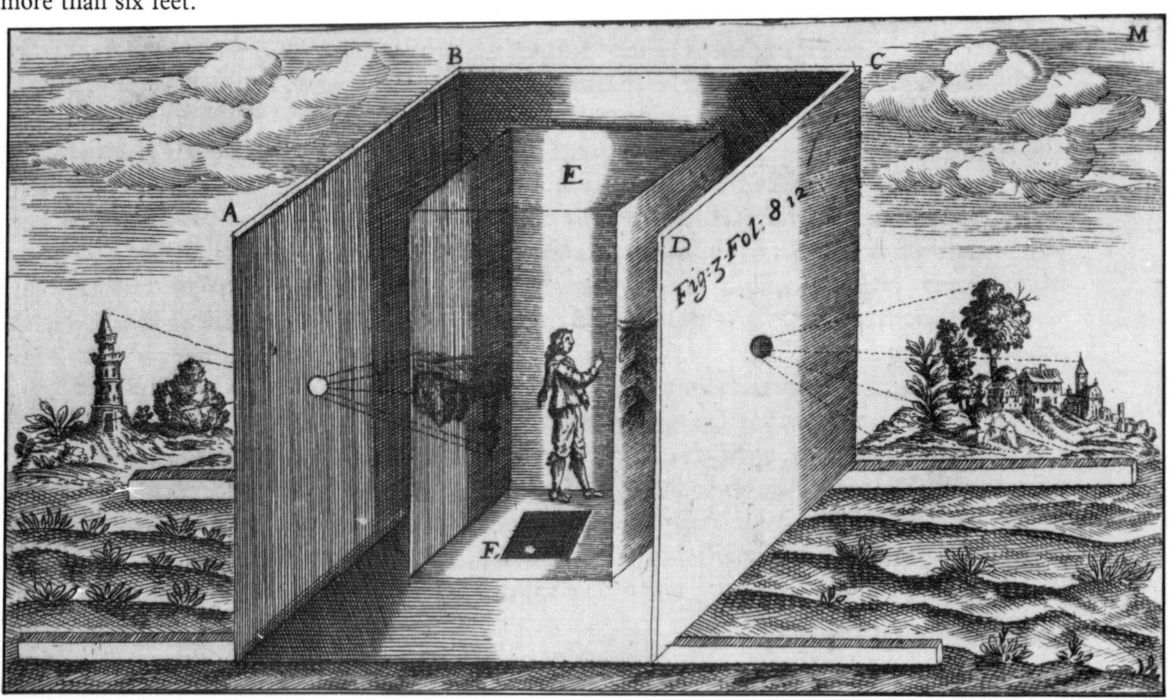

16. A double camera obscura from Kircher's *Ars Magna* (1646). It is unlikely that the room was twice the height of the man; the addition of the figure was merely a convention to show how the apparatus was used. The camera obscura may have been large enough to admit the head and shoulders. The distance from screen to screen was probably three or four feet, and between the two outer walls no more than six feet.

However, the apparatus is particularly interesting in that the illustration shows it to be two camera obscuras at 180° to each other. There is an outer box with a hole at two opposite sides, while inside is another box or frame covered with translucent paper. The draughtsman within is able to see an image on two sides of his little room. One half of Kircher's camera obscura is in principle very similar to that of Robert Hooke, and the finished drawings from both would be the correct way round. It is also said that Kircher made the first projection lantern.[16]

In 1642 a French mathematician, Pierre Herigone in his book *Supplementum Cursus Mathematici*, described a camera obscura contrived into a goblet.[17] A lens and mirror at the base of the stem projected an image which could be seen when the glass was filled with white wine—it was an ingenious device for observing others closely without their knowledge. A German schoolmaster, J C Kohlans, made a camera obscura in the shape of a book, which he called an 'opticum libellum'.[8] This may also have been a disguise, as was the goblet, or merely a decorative way of finishing off a camera obscura.

A modern author, J Marek of the Institute of Metallurgy in Prague, mentions two seventeenth-century astronomers

17. The goblet camera obscura described by Herigone and illustrated by Zahn (1685). The lens A projected an image which was reflected by the mirror f through the stem of the goblet onto the base of a glass cup insert at CD. The goblet lid with a lens at D served as a magnifier.

who used a special type of camera obscura to observe the interference patterns of light.[18] Balthasar Conrad, a professor at the Universities of Prague and Olomone, and Balthasar Melchoir Hanel both noticed 'rainbow effects' when they were making measurements of the sun's diameter by means of a camera obscura, but they attributed the effect to drops of water. At the time another professor at Prague, Ionnes Marcus Marci de Cronland, agreed with the existence of the interference effects, but disagreed as to the cause, saying that they were due to refraction in the observer's eye.

William Molyneux of Dublin had considerable experience in lens grinding and was the first to publish an English treatise on image formation in the camera obscura, both with and without a lens, and related to the eye.[19] He also referred to the use of a combination of lenses in order to produce telescopic effects, and gave rules for finding the focal length of lenses. He acquired a great number of mathematical and optical instruments, many of which remained in the family until the death of his son when more than six hundred pieces of apparatus were sold by auction. Among these was a camera obscura presumed to have belonged to William Molyneux since it is known that he bought a number of new instruments during his last few years.

In 1652 Père Jean-François Niceron, in a remarkable work on perspective and the drawing of curved surfaces,

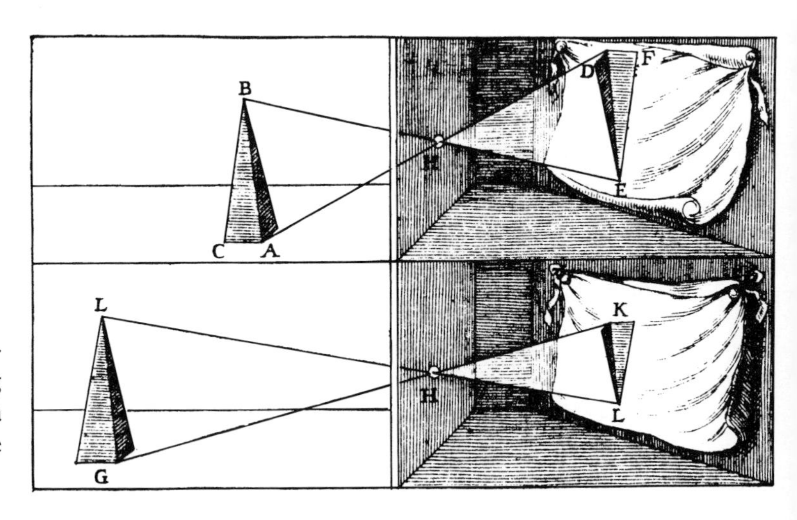

18. A drawing from Niceron's *Perspective Curieuse* (1652), demonstrating that the size of an image in a camera obscura is related to the distance of the object.

described the creation of images in a darkened room by a small hole in a shutter.[12,20] He enhanced this natural image by placing a long-focus convex lens in the hole, by means of which one could see the movement of figures that even a painter would be unable to reproduce. He also mentioned the inverted image, and the fact that it could be made upright by the adoption of lenses or mirrors. Niceron continued with an amusing piece of social comment. He observed that such a spectacle as the camera obscura could be seen at La Samaritaine, an area where ladies and gallants with too much time to spare forgathered. There were many sideshows, stalls and a camera obscura owned by charlatans who arranged outlandish scenes in its view. Their accomplices then took advantage of the darkened room to steal bags and purses from the fashionable customers who, captivated by the pictures, were often so naïve and unaware of the optical device as to believe it was all magic. So prevalent was the belief in astrology and the occult in the seventeenth century that many went to the camera obscura for predictions which were readily supplied by the quick-witted owners.

19. A combined camera obscura and study of the seventeenth century. Illustration from *La Dioptrique Oculaire* by Cherubin d'Orleans, 1671.

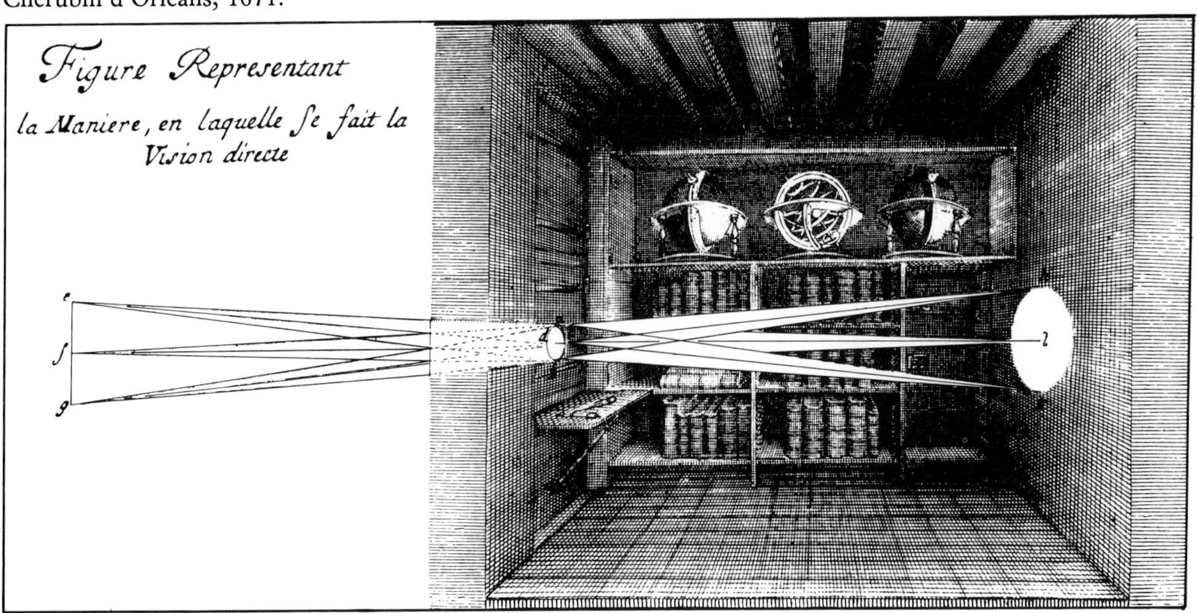

Cherubin d'Orleans in his *La Dioptrique Oculaire* (1671) provided an illustration of a darkened room with a projected image in which the letters show it to be upside down. The engraver did not make the room completely dark, so that it can be seen to contain many books, two globes and an armillary, representing the combined library and study of a gentleman of the period. Also, in a contemporary English translation of a work by Jean Leurechon, entitled *Récréations Mathématiques*, there is a description of a room-type camera obscura and a tent-like structure with a lens and image of a man who is standing outside the tent. This work was translated by William Oughtred, perhaps better known as the inventor of a slide rule, and for his suggestion that '×' should be used as a symbol for multiplication.

Christophoro Scheiner was a German monastic scholar who lectured at the Jesuit academy in Rome.[21] He was

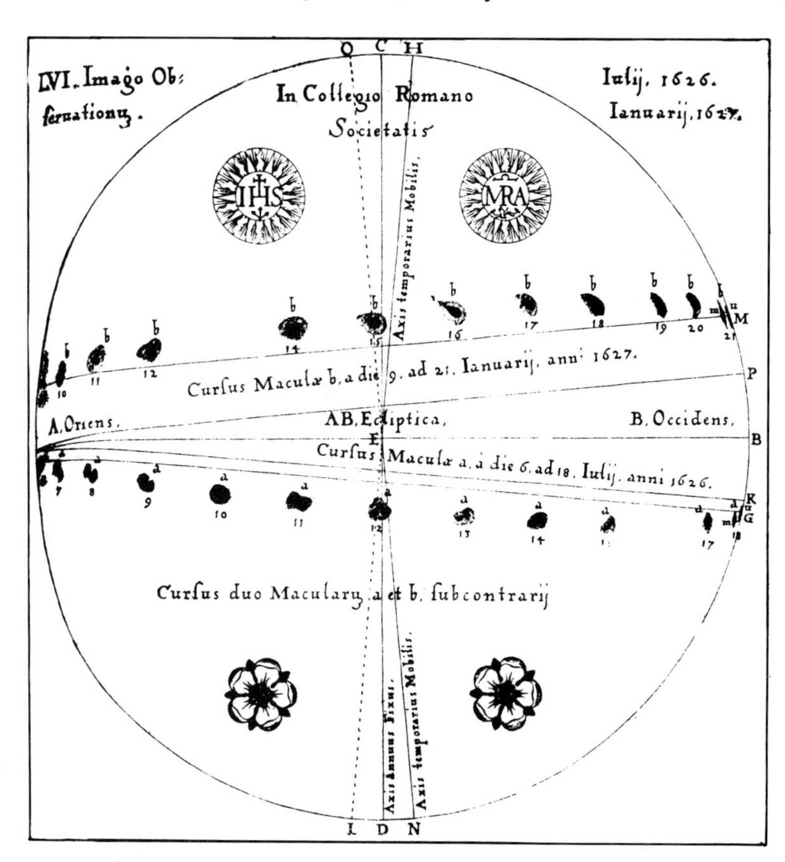

20. One of the many sunspot observations made by Christophoro Scheiner. From *Rosa Ursina Sive Sol* (1630).

21. An illustration from Scheiner's *Rosa Ursina Sive Sol* (1630) showing a camera obscura constructed between the eyepiece of a telescope and a screen. A cloth thrown over the framework made the image clear and bright, and by lifting the side it was possible to observe and measure sunspots. It was not necessary for the entire room to be darkened. This picture is part of a full-page illustration which, as the title says, consists of 'various methods for observing dark and light spots' of the sun. The sentence in the picture 'immission by multiple refraction' refers to projection with a telescope.

particularly interested in sunspots and was probably the first astronomer not to make his entire room into a camera obscura. He constructed a framework between the eyepiece of his telescope and the board which received the image of the Sun. He draped a black cloth over the framework, and made measurements of the sunspots by lifting one side only. It was not necessary for the windows in the room to be shuttered. Apart from astronomy, Scheiner was especially interested in vision and in his book *Rosa Ursine Sive Sol* (1630) he included a number of drawings to demonstrate by means of ray diagrams various defects of the eye which distort human vision. Each of these drawings was accompanied by a similar one of a camera obscura with a lens of the appropriate curvature and focus to produce the same ray diagram.

Another Jesuit scholar, Kaspar Schott, used the eye of an ox to demonstrate its similarity to a camera obscura by showing the image on the retina of the eye.[8] This was a delicate piece of work; the back of the eye had to be scraped so that the retina would be thin enough for the image to be

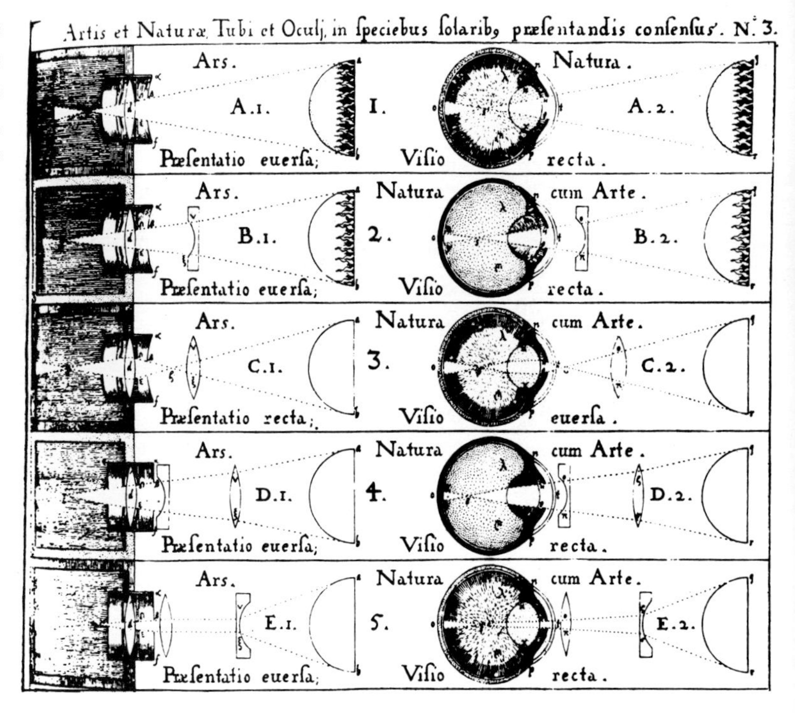

22. Ray diagrams demonstrating the similarities between the eye and the camera obscura. Each diagram shows the effect of adding concave and convex lenses to the normal image-forming lens. (From Scheiner 1630.)

seen. Schott also mentioned having heard of a small camera obscura which could be easily carried, and commented that there was no need for one's head to be inserted into the

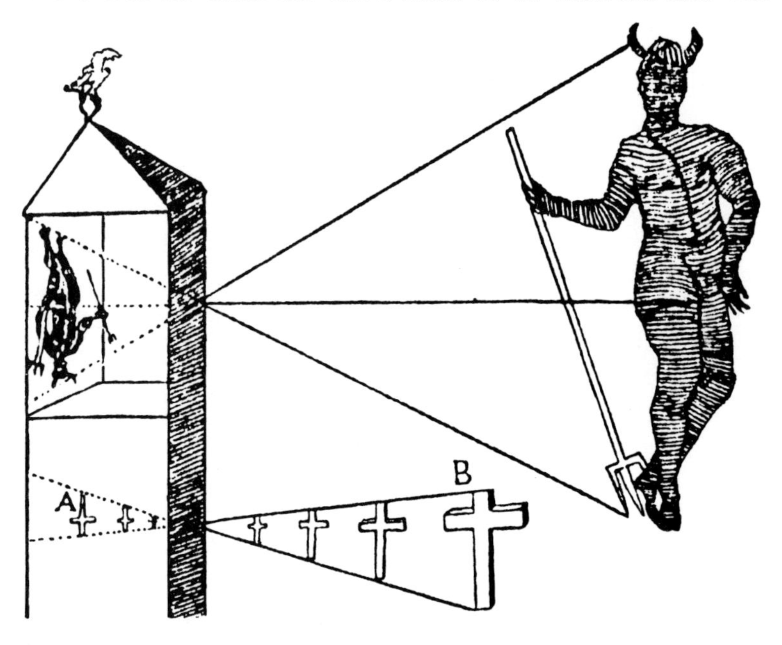

23. An actor and a camera obscura, from Scheiner's *Oculus* (1619).

24. A camera obscura from *Magia Universalis* by Gaspar Schott (1657).

apparatus. His design for a portable camera obscura consisted of two open-ended boxes which fitted one within the other. In the base of one he fixed a lens, and across a large hole in the base of the other he stretched a sheet of paper to form a screen. Sliding the boxes together provided a means of focusing the image. For making the image upright he suggested the use of two convex lenses at a suitable distance apart.

The scioptric ball, a lens in a swivel mount, was invented by Daniel Schwenter, professor of mathematics and oriental languages at Altdorf.[8.17] It consisted of a wooden sphere

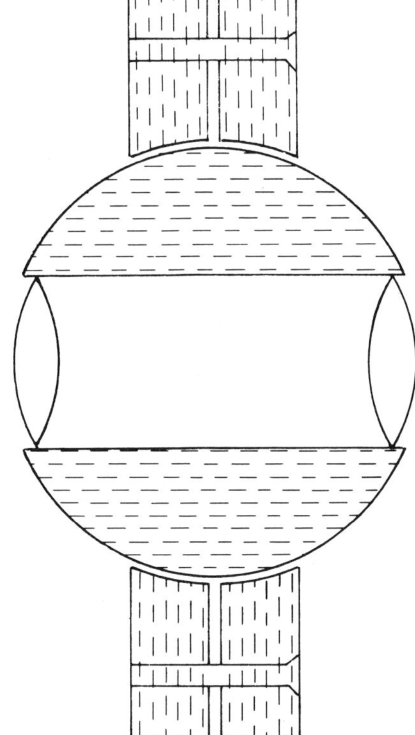

25. Diagram of the scioptric ball invented by Dr Daniel Schwenter in 1636. Two lenses were mounted in a wooden sphere held between two plates, and these gripped the ball sufficiently for it to be swivelled from side to side. When fitted to the shutter of a darkened room the scioptric could be turned to present images from either side of the view directly in front of the room. Scioptric lenses were used for making panorama drawings.

26. A scioptric lens made in the eighteenth century, with a mount of turned walnut.

with an axial hole into which was fitted a compound lens. The ball was held between two plates each with a hole slightly smaller in diameter than the axis of the sphere, in a manner similar to a ball and socket joint. When fitted into a shutter of a dark room, the ball was swivelled from side to side thus giving a continuous series of images of the outside scene. By this means it was possible to draw panoramic views, and Schwenter noted that such a view had been made of Nuremburg by Johann Hauer. Another professor of mathematics at Altdorf and one-time rector of the university, Johann Christopher Sturm, wrote a compendium of contemporary inventions and devices such as the telescope, airpumps and the diving bell, including a number of small portable reflex camera obscuras of box construction.[8,22,23] The image from the lens was reflected by a plane mirror at 45° so that it appeared at the open top of the box, across which was stretched a piece of oiled paper. The reflected image became upright but it was still reversed left to right. Oiling or waxing the paper made it more translucent and a hood at three sides of the paper shielded it from some extraneous light and thus helped to make the image appear a little brighter.

27. In order to study sunspots the scioptric ball was sometimes fitted with an adjustable mirror which reflected the sun's rays horizontally through the lens. A scioptric with a mirror was also adapted to the solar microscope, as was this eighteenth-century specimen from the Science Museum in London. The tube served to align and connect the microscope to the scioptric. Some astronomers removed the lenses and fitted a telescope in the ball which allowed easy adjustment of the telescope to follow the movements of the sun.

28. Interior of a camera obscura for solar observations, from Zahn (1685). A telescope was mounted in a scioptric ball with a rod firmly attached extending to the easel. The rod enabled the observer to adjust the angle of the telescope with the apparent movement of the sun and so maintain a constant position of the projected image. The scioptric ball, telescope and clamp are shown as individual items on the floor of the room. The height of the screen was adjusted by the screw pillars, and the angle of the screen relative to the telescope axis was adjusted presumably by the saw-tooth angle bracket, but exactly how it was fixed and held is not clear.

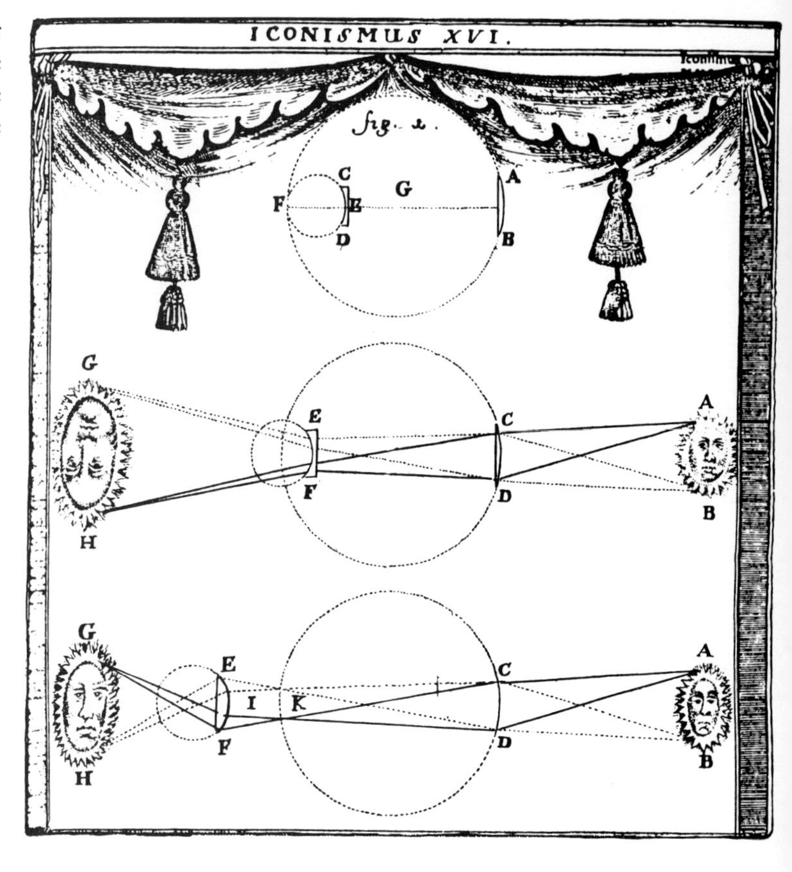

29. Ray diagrams from Zahn's *Oculus Artificialis* (1685), demonstrating the use of a concave lens to make a large image and a convex lens to erect the inverted image.

Johann Zahn in his book *Oculus Artificialis* (1685–86) also illustrated a number of box camera obscuras, one of which is of particular interest.[17] The glass screen receiving the image was not at a right angle to the lens axis, but at a slight angle down from the top to the back of the box. The mirror used was shown to be suitably adjusted for oblique viewing. This arrangement was much more comfortable for the draughtsman and allowed for a more efficient hood to shield off unwanted light. Focusing was achieved by sliding the lens in a tube fixed to the front of the camera obscura. Zahn illustrated the use of a second biconvex lens in order to correct the inverted image, and he also showed a telescopic effect by the addition of a negative lens, by which means he was able to produce enlarged images without the need for an unduly long extension to the camera obscura. It is in Zahn's book

30. A room camera obscura for observing sunspots. The sentence along the ray is in a poetic form of Latin which may be translated as: 'It shows the dark spots from high heaven.' (From Zahn 1685.)

that we find engravings of the goblet camera obscura described by Herigone, and a diagram of Schwenter's scioptric ball.

Appendix

The seventeenth-century antiquarian and gossip John Aubrey wrote: 'Old Goodwife Faldo (a Natif of Mortlak in Surrey) did know Dr. Dee, and told me that he did entertain the Polonian Ambassador at his howse in Mortlak, and dyed not long after; and that he shewed the Eclipse with a darke Roome to the said Ambassador.' From Aubrey's *Brief Lives* (1978) ed O L Dick (Harmondsworth: Penguin).

References

1. *Encyclopaedia Britannica* 1910
2. *British Journal of Photography* December 1857

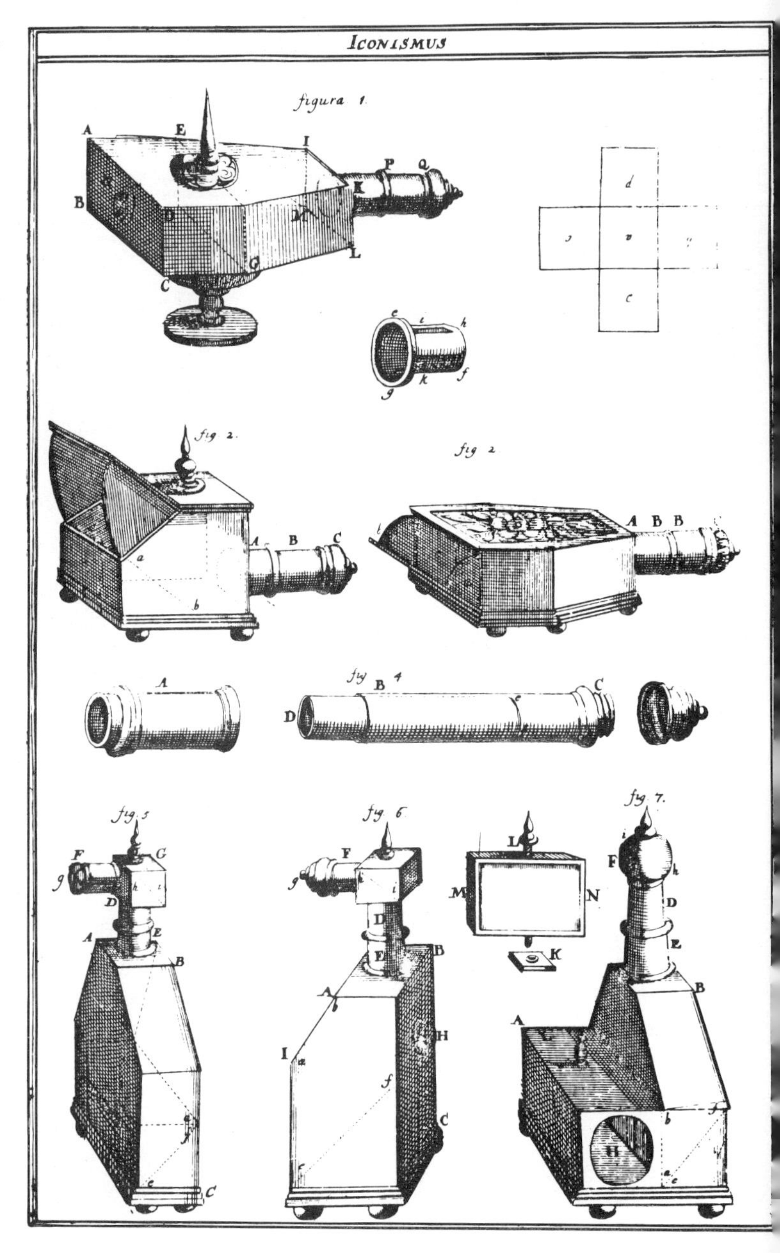

31. A collection of portable camera obscura designs from Zahn (1685). The long lens tubes were probably for close-up work and for telescopic lenses.

3. Waterhouse J 1901 *Photographic Journal*

4. Boyle, Robert 1669 *Of the Systematical and Cosmical Qualities of Things* (Oxford: Royal Society)

5. Whatton A B 1859 *Memoirs of the Rev. Jeremiah Horrox . . .* (London)

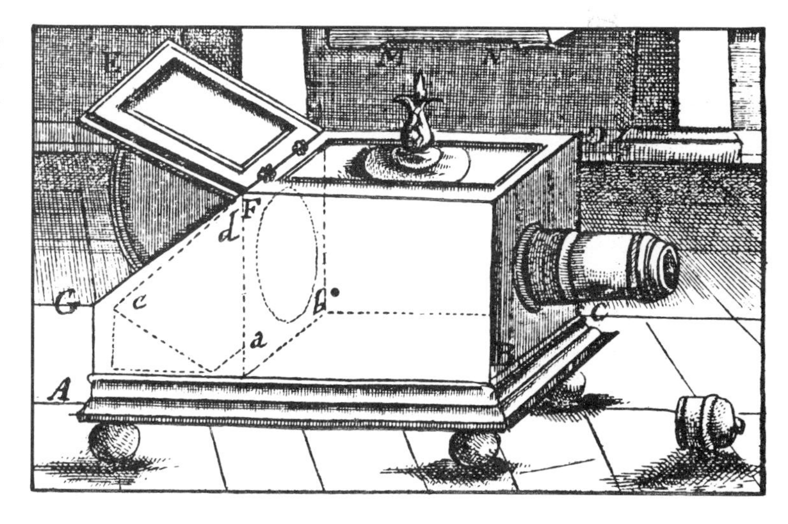

32. A reflex camera obscura with a screen at a comfortable angle for viewing. From Zahn's *Oculus Artificialis* (1685).

6. Waterhouse J 1909 *Photographic Journal*
7. *British Journal of Photography* March 1889
8. Gernsheim H 1969 *History of Photography* (London: Oxford University Press)
9. Pepys, Samuel 1666 *Diary* February
10. Tierie G 1932 *Cornelis Drebbel* (Amsterdam)
11. Bradbury S and Turner G L'E 1967 *Historical Aspects of Microscopy* (Cambridge: Heffer)
12. Potonniée G 1936 *History of the Discovery of Photography* (New York: Tennant and Ward)
13. *Encyclopaedia Britannica* 1973
14. Gage S H and Gage H P 1914 *Optic Projection* (New York)
15. Grigson G and Gibbs-Smith C H 1954 *People, Places and Things* vol. 3 (London: Waverley)
16. Priestley J 1772 *History of the Present State of Discoveries Relating to Vision, Light and Colours* (London)
17. Zahn J 1685–86 *Oculus Artificialis* . . . (Würzburg)
18. Marek J 1964 *Nature* vol. 201 January
19. Hoppen K T 1970 *Common Scientist in the 17th Century* (London: Routledge and Kegan-Paul)
20. Niceron P J-François 1652 *La Perspective Curieuse* (Paris)
21. Scheiner C 1630 *Rosa Ursina Sive Sol* (Rome: Bracciano)
22. Thorndike Lynn 1958–64 *History of Magic and Experimental Science* (New York)
23. Clay R S and Court T H 1932 *History of the Microscope* (London)

The Camera Obscura in Art

An instrument to take the draught or picture of any thing.
Robert Hooke

Drawing in perspective and the graphic representation of space in pictures had been mastered by Greek and Roman artists before the first century AD.[1] Villas in Pompeii and Rome often had wall paintings which depicted windows looking out onto natural landscapes; some of these pictures still remain and demonstrate how well the rules of perspective were appreciated. During the Dark Ages, however, there was a decline in cultural activities throughout Europe and many artists' skills were lost. It was not until the Renaissance that the ability to represent depth and space was regained.

During the process of this recovery artists devised many mechanical aids for drawing in perspective; for example, the architect Brunelleschi made a theoretical approach to the problem by producing a geometric formula which enabled painters to represent architecture accurately in their pictures. Alberti and Dürer both developed techniques and apparatus for drawing in perspective and foreshortening, but neither actually referred to the use of the camera obscura. In fact, there is no mention of its use by artists until the sixteenth century, when it is referred to as an aid for drawing.

During the next two hundred years the camera obscura became more widely known among artists, and during the eighteenth century many instrument makers were producing it and advertising it for sketching. The apparatus appears to have been a normal piece of equipment for travellers and amateur artists, but little is known of its use by professional artists of the period. Nevertheless, dictionaries and encyclopaedias frequently refer to Vermeer and Canaletto as having used the camera obscura; art historians have also made the same observation, and there is a considerable amount of literature on the subject.

Much of the conjecture surrounding Vermeer and the camera obscura is based on a supposed similarity between his rendering of certain objects and an out-of-focus image in a photograph, and some comparisons have been made. It is said that some of his viewpoints are as from a first-floor apartment and indicate the use of a room-type camera obscura. An example is the 'View of Delft', a wide panoramic scene in which it is supposed that the brush strokes resemble circles of confusion as produced by a lens†. It is also said that some of Vermeer's groups of figures lack cohesion resulting from the different viewpoints used, as though he set up a camera obscura for one figure and later used it in a different position for the second figure. These aspects of Vermeer's paintings and his relationships with other painters are discussed at great length by Fink and Wheelock.[5,6] However, in spite of all their investigations, factual evidence eludes these authors and unfortunately they do not even appear to agree on which of his pictures have circles of confusion, nor on the reasons for the slightly odd placement of figures.

Fascinating though these ideas are, their interest lies mainly in the argument, and even this wanes when one is not too sure whether the authors were aware of the difference between a camera obscura and a camera lucida, since so much is talked of the 'light' in Dutch paintings. Other assumptions have been based on Vermeer's connections with scientists. The scientist Balthasaar de Monconys may have introduced the camera obscura to him; it is known that he visited Vermeer with the intention of buying a painting, but apparently he was not successful. Anthony Van Leeuwenhoek, a cloth merchant and amateur scientist

†Also 'Girl with a Red Hat' (Pollack).

33. 'A View of Delft' by Vermeer, one of the paintings in which it is said that the artist imitated the appearance of the image in a camera obscura. The small points of white paint seen on the side of the barge (detail) are referred to by art historians as the optical way, or pointilles, or circles of confusion.

famous for the simple microscopes he made, was appointed by the Delft city council to act as executor to Vermeer's estate; it has been suggested that he may have shown the artist a camera obscura or even given him one, but there is no evidence that the two met socially, and a camera obscura was not listed among Vermeer's effects. One presumes that the estate was wound up carefully; Leeuwenhoek was a merchant and citizen of some standing, and Vermeer's widow was not too well provided for—she had eleven children.

Carl Chiarenze wrote: '. . . it is not unlikely that some Dutch artists used the camera obscura. There is, however, no known documentary evidence to show that they did.'[7] Swillens made many reconstructions of Vermeer's paintings, but it is said that he refused to see the necessity of supposing that the artist used any optical aids, either mirrors or a camera obscura.[4] In spite of making many suggestions as to the possibility of Vermeer's use of the camera obscura, Schwarz finally admitted that it 'cannot yet be established by any unimpeachable evidence or proof.'[3]

The use of the camera obscura by Canaletto has frequently been mooted and is supported by the quotation 'Canal taught the proper use of a camera ottica', which comes from a history of painting in Venice by Antonio Mario Zanetti (the younger) written in 1771, only three years after the death of Canaletto. Zanetti was librarian at the church of St Mark in Venice, and as well as being a connoisseur of art, he also made a number of etchings himself. The Correr Museum in Venice has a camera obscura with 'A. Canal' inscribed on the lid, but only wishful thinking can attribute it to the artist.[8] The book *Canaletto* by Bindmann and Puppi[9] illustrated this instrument, while Coke referred to a Venetian camera obscura from the workshop of Domenico Selva (dated about 1760).[7] The instruments appear to be similar but there is no mention of the Correr Museum or Canaletto in connection with the latter.

It is said that while walking about Venice, Canaletto frequently made notebook sketches of the city and it has recently been suggested that they were made with a camera

34. 'The Campanile after being struck by Lightning on 23rd April 1745' by Canaletto is an ink and wash study in which the perspective is obviously inaccurate. Although it is often alleged that the artist used a camera obscura for his sketches, it is most unlikely that he used one for this drawing unless he deliberately sketched part of it freehand, rather than by tracing an image. Reproduced by gracious permission of Her Majesty the Queen.

obscura. Many of these sketches have conflicting perspectives which could not appear in such an image, and it is unlikely that an artist would deliberately not accept true perspective when it is presented.[8] In any event, Canaletto would have been able to make notebook sketches much more quickly freehand than by tracing images; in addition, a camera obscura would have had to be supported.

Gioseffi constructed a camera obscura with which he made drawings so similar to those of Canaletto that he maintained that 'the drawings of the Quaderno are such as would have been produced using a camera obscura.'[10,11] Professor B A R Carter reviewed Gioseffi's book and was satisfied that for almost all the drawings Gioseffi was right in his conclusions. For his experiment Gioseffi made a small tent camera obscura which he used on a table similar to the design by Nollet. It is unfortunate that in his review Professor Carter confused the camera obscura owned by Sir Joshua Reynolds and that at the Correr Museum. Gioseffi's and Reynolds' camera obscuras are similar in principle; they both produce a correct image, whereas the one in the Correr Museum produces an upright image which is reversed from left to right. The instrument made by Gioseffi would have produced a bright and easily traced image, but it would have had to be supported on a table. One may still wonder whether Canaletto used such apparatus. However, that it was used by some artists is shown in the engraving by Costa (p49).

In his book *Canaletto and His Patrons* (1977), J G Links disputed the widely held belief that Canaletto used a camera obscura. Noting that since Canaletto claimed to have painted out of doors, which was an unusual practice at that time, it may have led his contemporaries to believe that he therefore needed a camera obscura.[12] Links suggested that it was Canaletto's ability to paint so realistically which has deceived many—from Algarotti to present-day historians—into believing that he used a camera obscura. In addition to this argument, Links explained that many of Canaletto's paintings of Venice included groups of buildings which cannot be seen from a single viewpoint; it is quite likely that they resulted from his experiments with the wide-angle view and his understanding of what his patron expected to see in the finished picture. Canaletto, as well as many other topographical artists of the period, may have used the camera obscura for a preliminary exploration of a scene in order to observe the composition and relationship of subjects within a frame. Perhaps finally one should consider the *complete* sentence from Zanetti, translated by Bindmann and Puppi:

> Canal taught the proper use of the camera ottica, and showed what defects can be introduced into a painting when its whole perspective arrangement is taken from what can be seen in the camera, particularly the colours of the atmosphere, and when one does not eliminate things offensive to the senses.[9]

On the back of a pen and ink drawing of Warwick Castle by Canaletto there is reputed to have been an inscription stating that it was made with a camera obscura, but unfortunately the drawing is now mounted. Perhaps one day an x-ray film will allay the doubts of art historians.[13]

Canaletto's nephew Bernardo Bellotto also became a master of perspective drawing.[14] After his apprenticeship he travelled, for a while became court painter at Dresden, and then went on to Vienna, Munich and finally to Warsaw, where he worked for the king and lived the last of his years. His groups of figures are more interesting and have more humour than those of Canaletto; he sometimes took his uncle's name and has been referred to as Canaletto the younger. The extraordinary accuracy of his town paintings made him famous in his own time, and in ours his series of Warsaw were used to guide the reconstruction after the last war. It has been said that he used a camera obscura, but Links argues that Bellotto, like his uncle, had such innate skill and assurance of vision that such an optical aid was quite unnecessary.[15]

After so much conjecture it is refreshing to discover those artists who left evidence that they actually used a camera obscura, or that they were familiar with the instrument.[16]

William Hogarth would have nothing to do with the camera obscura. He said that whatever he saw by way of subjects was more truly a picture than anything in a camera obscura, even to the extent that he 'grew so profane as to admire Nature beyond pictures and I confess sometimes objected to the divinity of even Raphael Urbin Corregio and Michael Angelo for which I have been severely treated.' Hogarth did not make a reference to the camera obscura in his final version of *The Analysis of Beauty*, but one can be found among the passages which were rejected by Hogarth and have been collected and edited by Burke.[17]

Nathaniel Dance exhibited a large landscape painting at the Royal Academy in London in 1794, upon which John Copley commented to his fellow academician Joseph Farrington that it was of the camera obscura kind. The painting has also been referred to as being in the style of the Dutch masters of landscape.[18] In his *Critical Guide to the Exhibition of the Royal Academy of 1796*, Anthony Pasquin commented on a painting: 'We think if Mr. Towne used a camera occasionally, he would correct his present manner of colouring.'[19]

A camera obscura once owned by Sir Joshua Reynolds is now in the Science Museum in South Kensington, London. It is of the type which, when collapsed and packed, resembles a very large book, although there is no direct evidence that Reynolds ever used it for drawing. Whilst on a tour of Holland, Reynolds commented on the paintings of Jan van der Heyden that they had a camera obscura quality, and in a letter to Edmund Burke he noted that 'Dutch pictures are a representation of nature, just as it is seen in a camera obscura.'[20] It is unlikely that Reynolds would have used one for a painting; in fact, in one of his lectures to the Royal Academy he suggested that any artist with powers of selection could make a superior representation of a scene to that of a camera obscura, in spite of its true picture of nature.[16]

An engraving by Gianfrancesco Costa of a canal with a village beyond shows a camera obscura in the lower left corner; it appears to be a small pyramid type on four legs,

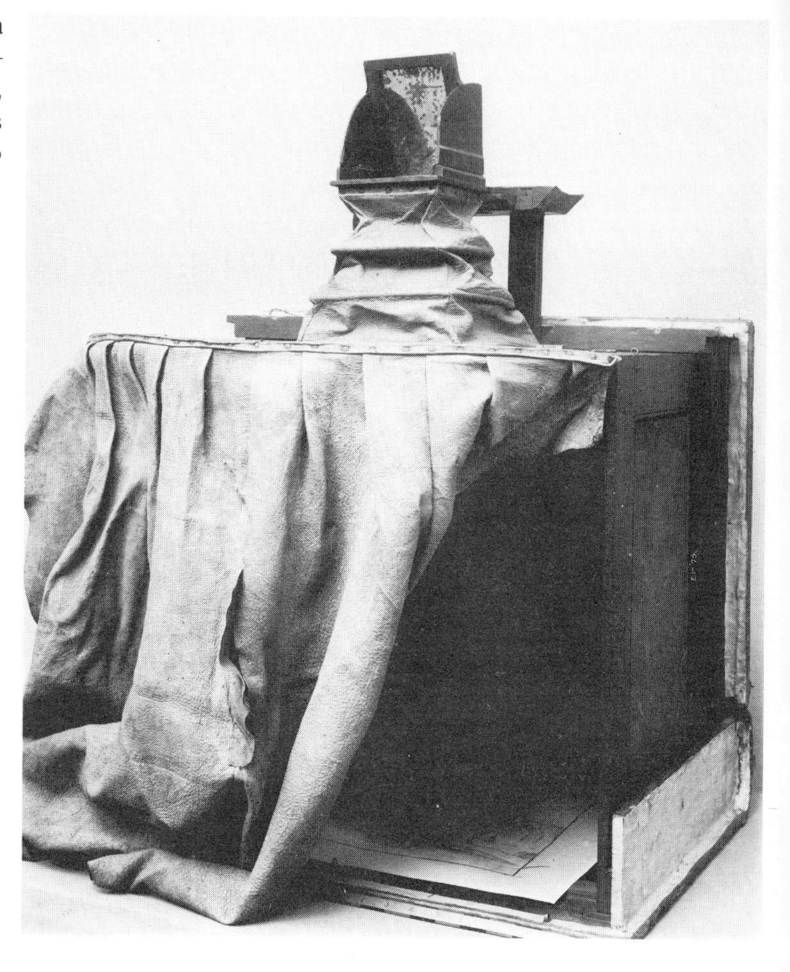

35. A book-form camera obscura owned by Sir Joshua Reynolds (1723–92), now in the Science Museum, London. The thick hide curtain was draped over the viewer's head to exclude as much light as possible.

attended by two people with an umbrella. The group seems to be almost a caricature, as though Costa was amused by the amateur artists. He was a pupil of Tiepolo, and although he has been described as an architect and painter, he is known mainly for his engravings.

Thomas Sandby, when he was employed at Windsor as court and estate artist, made a drawing of the castle and signed it 'Windsor from the Gossells, drawn in a camera T.S.' It is a panoramic view of the river and town with the castle in the centre. The drawing measures 12.5×58.2 cm, consisting of four pieces of paper pasted together, and is now in the library in Windsor Castle. Unfortunately the

36. Gianfrancesco Costa included a small tent camera obscura on a stand, similar to that of Abbe Nollet (p76), in this engraving of a canal winding between houses and a church (Veduta del Canale verso la Chiesa della Mira). Engraved in about 1750.

camera obscura used by Sandby has been lost, but one can surmise from the drawing that the picture area may have been about 15 × 15 cm, or a little larger than a single sheet of the drawing, and that the lens had a focal length of about 25 or 30 cm. Sandby made a drawing of the Gothic gallery at Strawberry Hill and Horace Walpole wrote on it 'T. Sandby drew this by the eye in two days and a half.'[18] It is possible that 'by the eye' meant not with a camera obscura. Walpole himself possessed a camera obscura made by William Storer. On the first day of the sale of Thomas Sandby's effects a large camera obscura was an item in the catalogue.[21]

37. 'Windsor from the Gossells, drawn in a camera T.S.' A drawing made by Thomas Sandby (1723–98) on four pieces of paper, making a total length of 58 cm. Reproduced by gracious permission of Her Majesty the Queen.

Paul Sandby, brother of Thomas, made a water-colour painting of Rosslyn Castle, in the foreground of which is a lady using a camera obscura.[16] The instrument appears to be a box reflex type mounted on a stand consisting of a pillar with three curved feet, of the type used for small tables of the period. It is difficult to gauge the exact size of the drawing area of the apparatus, but it was probably about 20 × 15 cm, and the focal length of the lens perhaps 30 or 40 cm. The painting is in the Paul Mellon Collection and was on view at the Royal Academy in London in 1972, although the catalogue of the exhibition unfortunately refers to the instrument as a camera lucida.

The National Gallery in Washington, DC, has a charming portrait painting of three children by Charles-Amédée-Philippe van Loo. The picture has the fascinating *trompe l'oeil* device of a painted frame of rounded stonework

38. 'The Magic Lantern' by Charles-Amédée-Philippe van Loo (1719–95), now at the National Gallery of Art, Washington. It is almost certain that the boy is holding a camera obscura, which the artist was more likely to have in his studio than a magic lantern. Gernsheim referred to this picture as a 'mid-eighteenth-century box camera obscura'.

through which the children can be seen. A small child has her hand on the frame and a boy holds a camera obscura, which is on the viewer's side of the frame. We see only the front half of the box, which is made of wood and is quite plain except for a lens mounted on the front panel. The picture is entitled 'The Magic Lantern', but the construction of wood and the simplicity of the box, which has no slot for a slide, suggest very strongly that it is a camera obscura. There is obviously some doubt and confusion about the title of the picture. A gallery catalogue says: '. . . the figures are enframed by a circular wreath, thus simulating a scene viewed through the lens of the closed box—a rare example of a visual pun.' Such a box for looking at pictures through a lens or small hole is usually referred to as a 'show box'.

An eighteenth-century snuff mill on Clifton Down in Bristol suffered a disastrous fire soon after it was built and it became a dilapidated ruin.[22] William West rented the old mill in 1828 and after some rebuilding he converted it to the 'Clifton Observatory' by installing telescopes and a few other optical instruments. Later he added a room-type camera obscura which is still operating and open to the public. West was a member of the Bristol School of painters and often exhibited at the Royal Academy, but he does not appear to have used the camera obscura as an aid for drawing or making sketches for his paintings. He had many other interests and was at one time very much involved with an idea for a bridge across the Avon Gorge at Bristol, but Brunel tactfully took over the job.

A camera obscura appears in a portrait painting of the court painter Joachim Frans Beich by Georges Desmarées. Beich worked for some time for Max Emmanuel of Bavaria when he produced two large commemorative paintings of a victory by Bavaria over the Turks. The International Museum of Photography in Rochester, New York, has a mezzotint copy by Johann Gottfried Haid of the Beich portrait.

Thomas Daniell and his nephew William were travelling artists who often used a camera obscura,[23] and they spent several years in India making drawings and paintings of

39. 'Temple near Bangalore (Mysore) May 1 1792', a drawing by Thomas or William Daniell in which the pencil strokes have the appearance of a tracing. Each line is of uniform thickness and has a positive beginning and end. The cornice of the near temple appears to be elongated to the right. This kind of distortion occurs at the edge of an image, and although no proof is available, these characteristics suggest that this drawing was made in a camera obscura. Mrs Mildred Archer of the India Office Library is confident that the numerous drawings made by the Daniells were achieved by the use of a camera obscura on many more occasions than those mentioned in their diary. Reproduced by permission of the Director of the India Office Library and Records.

palaces, forts and other antiquities. William kept a diary in which there are many references to their work and the camera obscura. For brevity or perhaps familiarity he refers to his uncle as 'Un'.

> 11 Nov. 1788. After breakfast I went to the Gola & made two drawings of it in the Camera Obscura.

> 25 Jan. 1789. Left Agra abt. 6 o'c and arrived at our ground at Secundra (dist. from Agra—near 9 miles) abt. 8 o'c. Un[cle] employed the whole day drawing the Gate, in the camera, leading to the Tomb of Akbar.
> 18 Feb. 1789. We breakfasted very early & spent the day at Jammaigh Musjid, built by Shah Jehan. Un employed the camera, took a view of the Mosque with the Minorets.
> 28 Aug. 1789. Un dead coloured the View on a half length that I sketched yesterday, myself dead colouring the Tage Mahl. Mr. Bellas called on us in the morning & was so good as to undertake the repairing of the Camera obscura for us.

At the Conservatoire National des Arts et Métiers in Paris there is a camera obscura used by Louis Jacques Mande Daguerre, an artist who was one of the three inventors of photography.[24] Before investigating the photographic process he established an entertainment show in Paris called the 'Diorama'. This was an exhibition of large translucent paintings illuminated from the rear by lamps which could be made bright or dim and moved about in order to light up different areas of the paintings, by which means it was possible to change the mood and effect of a scene. Daguerre's painting 'The Ruins of Holyrood Chapel', now hanging in the Walker Art Gallery in Liverpool, is very suggestive of a camera obscura image. The strong lighting effect is similar to that in many paintings by Joseph Wright of Derby, of whom Klingender commented, 'Joseph Wright on the other hand, used the half light in a manner that is akin to the camera obscura of which Darwin wrote.'[25]

The increased light and colour contrast of the camera obscura image was also noticed by Ruskin: 'I have often been struck, when looking at the image of a camera-obscura on a dark day, with the exact resemblance it bore to one of the finest pictures of the old master, all the foliage coming dark against the sky, and nothing seen in its mass but here and there the isolated light of a silvery stem or an unusually illumined cluster of leafage.'[16] Ruskin's opinions on the old masters were very staid, and it would be very interesting to know what comments he would have made about some of the recently cleaned paintings at the National Gallery. It is

40. 'The Ruins of Holyrood Chapel' by Louis J M Daguerre (1787–1851). The intense light and dark of this painting is similar to a camera obscura image of a sunlit scene. The effect of brilliant highlights and rich shadows was noticed by Ruskin and is evident in many paintings by Joseph Wright. It is likely that Daguerre observed this ruin in sunlight with a camera obscura and made the painting as a moonlit scene.

not known whether he ever used a camera obscura other than for observation.

In the mid-eighteenth century the Venetian Count Francesco Algarotti was appointed art adviser to Frederick the Great of Prussia and Augustus III of Saxony.[16] As a patron he befriended many artists, including Canaletto and Tiepolo, and as a writer he attempted to break down the strong prejudice to new ideas and the cultural isolation of Italy at that time. He frequently said that artists should use the camera obscura and that it should be as much part of their equipment as the telescope and microscope were for the astronomer and biologist. As an example, Algarotti referred to Guiseppi Maria Crespi, also known as Spagnoletto of Bologna, and that the wonderful effect in his pictures was due to his use of a camera obscura. It is interesting to note that he suggested that the camera obscura should be used as a means for observing, rather than as a tool for direct drawing. This points to a difference of approach between the professional artist and the scientist.

It is likely that the camera obscura was used for book illustration to a much greater extent than we actually know; for instance, drawings from the instrument may have been used as a guide for an engraver to cut his plate. Afterwards the drawings would have been discarded as of no further use, and consequently we have no records. We do know, however, that William Cheselden, surgeon to Queen Caroline, used a camera obscura to illustrate his book *Osteographia, or the Anatomy of the Bones.*[26] The title page has an engraving showing a very large box camera obscura on two trestles and before it hangs a skeleton. It seems the draughtsman found it easier to draw objects the right way up—the skeleton hangs upside down! In his introduction to the book Cheselden referred to the task of making the illustrations:

41. A very large but simple box-type camera obscura used by Cheselden to make the illustrations for his work on bones, *Osteographia* (1753). The skeleton may have been upside down because Cheselden required a particular aspect of the bone structure, or maybe because the draughtsman found it easier to trace an image which was the right way up.

> Then we proceeded to others, measuring every part as exactly as we could, but we soon found it impossible this way; upon which I contrived (what I had long before meditated) a convenient camera obscura to draw in, with which we corrected some of the few designs already made, throwing away others which we had before approved of, and finished the rest with more accuracy and less labour, doing in this way in a few minutes more than could be done without in many hours, I might say in days.

42. 'The Lower Arch at Maktar, Tunisia', a drawing made by James Bruce while he was British Vice-Consul in Algiers. The eye level horizon of the picture appears to be at the base of the pillars, probably six or seven feet above the ground. This would be about the height of the mirror in the camera obscura used by Bruce. This reproduction was made from a lithograph in Playfair's book published in 1877, a little more than 100 years after Bruce made the drawing.

As if to justify his use of a camera obscura Cheselden made the point that when artists have occasion to draw straight lines or circles they do not disdain to use a ruler or compass. Edward Dodwell made the illustrations for his book *A Classical and Topographical Tour through Greece* with a camera obscura, and Robert Hooke recommended it specifically for book illustration.[27]

During his term of office as British Consul in Algiers in 1763, James Bruce went on a tour of exploration to discover the source of the Nile.[28] During this journey and other excursions in North Africa he made a large number of drawings of the ancient buildings and Roman remains with a camera obscura, and on his return to England many of the pictures were presented to King George III. Bruce, a strong, powerful man, was accomplished in many respects. His book *Travels to Discover the Source of the Nile* (1790) recounted a number of exciting adventures such as a stormy shipwreck, being stripped naked and beaten by Arabs who thought he was a Turk and, later, administering medical aid to the ladies of a harem. When in London he spent an evening with Dr Burney whose playful daughters referred to him as the Man Mountain; Fanny Burney wrote that he was 'the tallest man you ever saw in your life—at least gratis.'

Before leaving England for Algeria Bruce asked Messrs Nairn and Blunt, instrument makers opposite the Royal Exchange in London, to make a camera obscura to his own design. It was hexagonal, about six feet across, with a conical top like a summerhouse, in which a draughtsman could sit and work at his drawing on a table. Bruce commented that such an instrument allowed a person of only moderate skill in drawing to execute as much in one hour as a prope draughtsman could in seven. The whole apparatus wa separated into two parts and by means of hinges could b folded to a compact parcel which was not 'heavy, cumbe som or inconvenient'. When collapsed it looked like a 'hug folio book, about four feet long and ten inches thick.' In a account of his travels Bruce described the camera obscur and noted: 'There is now, I see, one carried as a show abou the streets, of nearly the same dimensions, called Delineator, made on the same principles, and seems to be a exact imitation of mine.' On arrival at Algiers, Bruce foun the camera obscura so effective that he ordered a smalle one, probably a box type, from Italy which 'though negli gently and ignorantly made, did me this good service, that i enabled me to save my larger and more perfect one in m unfortunate shipwreck at Bengazi.'

About a hundred years later, Lieutenant Colonel R I Playfair retraced the journeys made by Bruce and in 187 published *Travels in the Footsteps of Bruce in Algeria an Tunis.*[29] A footnote on page 4 stated that the camera obscur used by Bruce was then at Kinnaird in Scotland, but afte many unsuccessful enquiries made by the present author, i must now be presumed lost. Gernsheim suggests tha Goethe's reference to 'an Englishman who always took a camera obscura on his travels' may have meant Jame Bruce.[27] In the museum at Weimar in East Germany ther are two instruments which are thought to have belonged t Goethe, who was especially interested in the science of ligh and colour.

According to Scharf, a very good description of how t construct a camera obscura, how much it should cost and it use, appeared in a book published in 1732 by J Peele entitle *The Art of Drawing and Painting in Water Colours.*[16]

Peter Suhr, a lithographer, made sketches of Hamburg with a camera obscura before and after the great fire of 1842. This early example of work for architectural preservation is also mentioned by Coke.[7]

Perhaps one of the most fascinating examples of the use of the camera obscura in art was that made by the pottery manufacturer Josiah Wedgwood for the famous Russian service.[30-32] The Empress Catherine had requested a cream ware service for 'every purpose of the table' and each piece was to show a different view of British scenery. The service was to comprise some 952 pieces, depicting 1244 different scenes. Besides being a very fine artist and craftsman,

43. A plate from the 'Russian service' of tableware made by Josiah Wedgwood for the Empress Catherine, who asked for a different English view on every piece. Many of the drawings of country houses were made by artists using camera obscuras. The painting on this plate is of the lake at West Wycombe House where the grounds, including the lake, were landscaped by Humphrey Repton. The frog device was chosen by the Empress to represent the Palace Kekerekeksinsk, Frog Marsh. This plate, now at the Victoria and Albert Museum in London, was probably a spare or trial piece; most of the original service is now at the Little Hermitage, Leningrad. Reproduced by gracious permission of Her Majesty the Queen.

Wedgwood was also a clever businessman with considerable managerial skills, but nevertheless even he was at first a little daunted by Catherine's request. Eliza Meteyard quotes him as saying 'all the gardens in England would scarcely furnish subjects sufficient.' On receiving the order from the Russian Consul Wedgwood wrote to his London partner, Bentley, for a camera obscura and he despatched artists, some with camera obscuras, to take views all over the country. Still fearful of not having enough material, however, he also ransacked print shops and advertised for gentlemen to send him pictures of their houses. In a letter to Bentley he referred to a young artist neighbour, Mr Stringer: 'Do you think, it would be worth while to ingage Stringer for a few months to paint and instruct our hands in London? Upon this plan I would bring him up to London, have a camera obscura with us, & take a 100 views upon the road.'

The Wedgwood Museum at Barlaston, Stoke on Trent, has no information about the camera obscura Josiah ordered, and all we know of it is again from Meteyard: 'The camera obscura Bentley procured folded up and took large sized views.' The order for the Russian service came when trade was low and, although proud to have received the request, Wedgwood was a little concerned about payment. He need not have worried though, for the Empress promptly paid 16 406 roubles 43 kopeks (about £2700) for the service. Wedgwood's expenses were very high, however, and he made only a small profit of a few hundred pounds, but he put the entire service on show in London for a month before sending it to Russia. The service became a fashionable exhibition, and through it Wedgwood gained much prestige and his business improved as a result. Such a large number of landscape pictures on view at one time also stimulated an interest in topographical drawing and painting which may have contributed to the popularity of the camera obscura in the nineteenth century.

During the eighteenth and nineteenth centuries many instrument makers advertised their camera obscuras as being suitable for drawing. The Science Museum in London has a box-type instrument of the late-eighteenth century which is inscribed by a stamp or stencil 'Jones (Artist) London' within a circular announcement 'By His Majesty's Special Appointment'. These words are on the lens pull-out, and a second pull-out portion bears the address 'No. 4 Wells Street, Oxford Street'. All the letters are Roman except for the word 'Artist', which is in script and enclosed by brackets. It has been suggested that Jones was an artist, but it is more likely that since there must have been a number of instrument makers named Jones, the word 'Artist' refers to that particular model of camera obscura. (A further note about this instrument appears in the chapter on the eighteenth century.)

Founded in 1850, the well known firm of Negretti and Zambra suffered considerable losses at their London office during the blitz of the second world war, but Mr Zambra fortunately kept a few old catalogues at his home and they are now preserved by the firm at their new premises near Amersham. The catalogue of 1865 listed several camera obscuras for drawing and a 'Cosmorama, or camera obscura for gardens', as well as prisms and lenses 'in brass mountings with sliding adjustment for constructing garden

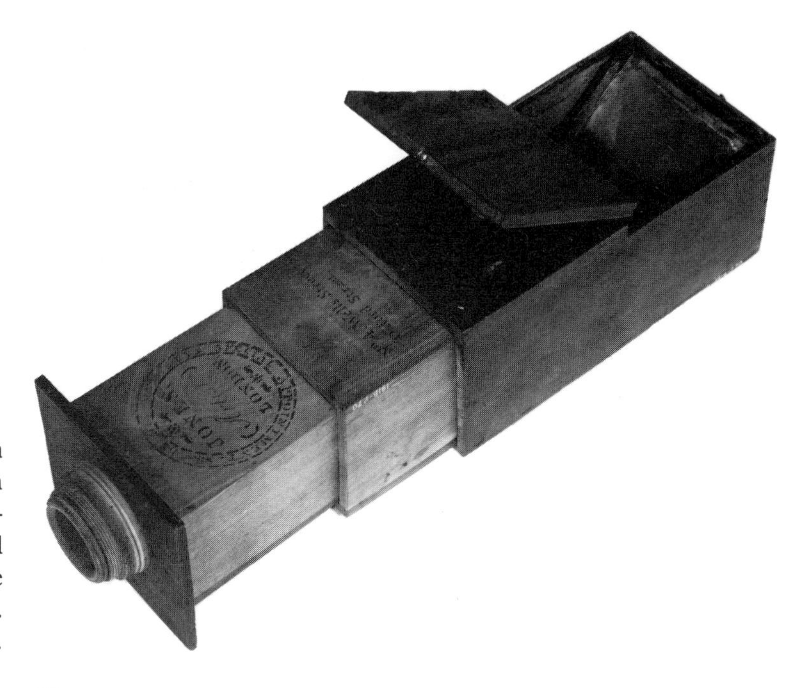

44. The 'Artist' reflex camera obscura made by Jones, London, in about 1790. The draw boxes for focusing are pushed in an out by hand, and there are no side pieces to shield the ground glass from extraneous light. The picture area is about 6×5 inches.

cameras.' The 1886 catalogue listed a 'Draughtsman's camera obscura (chambre noire) for sketching 21s, £1 15s 0d, £2 2s 0d', and in the margin is a handwritten note about a camera obscura made specially for 'C.H. at Cardiff May/June 1899'. The prices are also handwritten in a code which was never written down, although it was in use until about 1950. Mr Day, an employee of Negretti and Zambra was able to remember some of the code and estimated the cost to C.H. in Cardiff to be about £21. It is thought by some of the older members of the firm that the production of camera obscuras ceased after the 1914–18 war. During the 1930s, however, Negretti and Zambra made and installed a special camera obscura which will be mentioned later, but it was only for viewing and not drawing.

Martin Hardie who died in 1952 was for many years an artist employed by the Victoria and Albert Museum in London.[19] He borrowed Joshua Reynolds' camera obscura from the Science Museum and used it to make a drawing or, rather, an outline sketch of the nearby Imperial Institute. It took him about fifteen minutes, but he stressed that although it was merely a tracing, it had a precision of perspective which would have taken hours to draw free hand. He compared this brief experiment with the Daniell's method of working and suggested that William may have made the drawings in the camera obscura while his uncle embellished them with shadows, figures and anecdotes. This may have happened occasionally, but the Daniell diary frequently refers to both uncle and nephew using the camera obscura.

William Mason in his quaint little guide book *The English Garden,* written early in the nineteenth century, includes a footnote referring to a Claude glass as 'perhaps the best and most convenient substitute to a camera obscura.'[33] The Claude glass, named after Claude Lorraine the French land scape painter, was a convex mirror but instead of being silvered, the reflecting surface was painted black which made the glass reflect darkly, subdueing harsh colours and lights. Although dull, the reflection was gentle and dim suggesting a 'sublime mood', an aesthetic appreciation which was very much in vogue towards the end of the eighteenth century. The Claude glass cannot be considered as a drawing aid other than in presenting a dark reflection of a scene in miniature.

Today graphic artists and designers use a camera obscura to make a larger or smaller copy of an original which is not of a suitable size for a layout or design. The apparatus is now called a projector, and its function is similar to an

45. A megascope, from *The Forces of Nature* by A Guillemin, a type of camera obscura which made large images of small objects. The lens would have been about 30 inches focal length in order to make an enlarged picture within the space of a room. A mirror reflected extra light to the front of the object. Illustrations of the megascope appeared during the eighteenth and nineteenth centuries, although in the seventeenth century Risner had referred to the use of a camera obscura for enlarging drawings and prints.

eighteenth-century variant of the camera obscura called a 'megascope' which was used for making enlarged drawings of small objects or engravings. Messrs Grant & Co of Walton-on-Thames produce drawing office projectors capable of reducing or enlarging a picture to a ratio of 3:1 and 7:1. Similar instruments are also made by Messrs Halco-Sunbury & Co at Staines, and no doubt by other manufacturers too. Another example is the 'Variograph' made by Rudolf and August Rost of Vienna, which is a more versatile instrument although the reduction and enlargement ratios are about the same. As the name suggests, it is possible to adjust the scale of either the x or y axis of a graph or design by tilting the copy board and the drawing surface. This function is sometimes required in engineering and textile design to allow for stretch or shrinkage. The lens of the variograph is mounted on a sliding panel to enable large originals to be copied in portions; this feature obviates the need for a special type of wide-angle lens which is complex and costly.

In 1969 Philip Rawson, in his book on drawing, mentioned several mechanical devices to aid the draughtsman: 'The camera obscura is another machine which has been widely used by artists to help them in tracing accurate perspectives.'[34] He referred to Zanetti and Canaletto, and finally commented: 'It [the camera obscura] is still used in technology, and by fashionable portrait painters and other artists to help them in capturing likenesses.' In his book *Birth and Rebirth of Pictorial Space* (1957) John White referred to Vermeer and made interesting comments on perspective in paintings.[35] He said that during the seventeenth century some artists employed an artificial perspective whilst others experimented with a synthetic system, and that the camera obscura had caused much excitement. He referred to the first as rectilinear and the second as employing curved lines. One may wonder whether a perspective of curved lines resulted from, or was suggested by, either barrel or pin-cushion distortion (aberrations in an image caused when an aperture is placed in front of or behind a lens). In his book *Optics, Painting and Photography*

(1970) M H Pirenne referred to this distortion and the effect of wide-angle projection onto a flat surface at right angles to the lens axis, and also mentioned distortion of the peripheral image in the human eye.[36]

Many art historians have surmised that the apparatus associated with Alberti was a show box for the display of paintings. Kenneth Clark in *Landscape into Art* (1949)[37] referred to Alberti's having a sort of camera obscura, the images of which he [Alberti] called 'miracles of painting'†. Clark suggested that the realistic background of Italian paintings owed something to Alberti's 'magic box' and the varied panoramas it produced. When writing about the atmospheric effects in paintings of the Dutch masters, Clark referred to the camera lucida and that it was 'an artist's habitual companion'. This comment is somewhat confusing in view of the concern by many authors about the use of the camera obscura by Vermeer and other Dutch painters. Furthermore, the camera lucida was not invented until 1806 by W H Wollaston, and it is well known that it is not at all an easy instrument to use.

Many dictionaries and encyclopaedias of art include entries on the camera obscura, some of which are useful, although many are inadequate. It is frequently stated that the instrument was invented by Porta in the sixteenth century, and that one had been used by Canaletto. Descriptions of the instrument vary from constructional details to brief notes about collections of lenses and a mirror; sometimes the camera lucida is mentioned as being a better or more sophisticated instrument. Occasionally the colours in camera obscura images are said to be over-bright and not 'as in nature', and the perspective is often criticised. In the eighteenth century Georg Busch referred to the image in the camera obscura as being always exact in detail but distorted by perspective.[39]

†The term 'camera obscura' was frequently misused by itinerant showmen and street hawkers during the eighteenth century. A multiple show box or peepshow labelled 'Grand Camera Obscura' is the subject of a mezzotint reproduced by Peter Pollack in *A Picture History of Photography.*[38]

When questioned about the camera obscura, art teachers and historians today generally tend to be vague and uncertain about the instrument, often confusing it with other mechanical devices for drawing or copying, such as a framed grid or sheet of glass and sight stick, as illustrated by Dürer. Most people in the art world declare that the camera obscura gives a distorted view, but they are usually unable to explain why. Since our visual perception of the world is

46. A pen and brush drawing entitled 'Camera Obscura' by Jurriaen Andriessen, dated about 1800. From *A Picture History of Photography* by Peter Pollack; the drawing is at the Rijksprentenkabinet, Amsterdam.

tempered by preconceived and pre-learned information it is difficult to refute this comment. Gombrich suggests there is a schema or mental image which we relate to a situation or set of circumstances.[40] However much faith one may have in the objective geometry of an image made by a lens, the world of art is subjective and, in the final analysis, the appreciation of a picture is a personal judgment. Furthermore, it must be accepted that when a simple lens is used in a camera obscura the image is distorted, and curved rather than straight lines are produced at the edges of the image area. This fault would usually be corrected intuitively by an artist, but a painting at the Tate Gallery shows buildings on the extreme right and left which lean towards the centre (exhibit number 374, 'Column of St. Mark, Venice' by Richard Parkes Bonington)†. A relevant comment was made by James Bruce who, after describing his camera obscura, said,

> It is true, this instrument has a fundamental defect in the laws of optics; but this is obvious, and known unavoidably to exist; and he must be a very ordinary genius indeed, that cannot apply the necessary correction, with little trouble, and in a very short time.

To end this chapter it is appropriate to refer to a publication of 1658, *Graphice, or the Most Excellent Art of Painting*, in which a tent-form camera obscura was described.[41] It was suggested that the front lens of a telescope could be used, 'to which apply a long prospective trunck, with a convex glasse; fitted to the said hole, and the concave taken out at the other end. The tent may be turned round to design the whole aspect of the place.' Whilst recommending the camera obscura for topographical purposes the writer says that for making 'landskips hereby were too illiberal.'

†A 'large camera obscura' was listed in the sale of effects of Bonington's father, August 1817. (From a newspaper cutting at the Castle Museum, Nottingham.)

Appendix

The correct orientation of a drawing was important for topographical and scientific artists such as the Daniells and Cheselden. The simple camera obscura as used by Cheselden gave a correct image but it was upside down, and therefore difficult to draw. Although the reflex camera obscura produced an erect image it was reversed left to right. However, if one looks at the image made by a reflex camera obscura from inside the instrument the image is correctly orientated. This observation gave rise to the tent, pyramid and book-form camera obscuras which were used by the artist inside the apparatus or partially so.

The catalogue of the effects of Dr Thomas Monro referred to a 'camera obscura, with transparencies beautifully painted on glass, by Gainsborough'. This was Gainsborough's show box. Dr Monro was a connoisseur and the patron of a group of young artists which included Turner, Girtin and Cotman.

Nigel Konstam writing in *The Artist* (January 1980) stated: 'It seems to me highly improbable that Vermeer used the camera obscura. Wilensky suggested that he may have used two mirrors, and I believe that I have proved beyond reasonable doubt, that this is so.'

References

1. Grant M 1960 *The World of Rome* (London: Weidenfeld and Nicolson)
2. Gowing L 1970 *Vermeer* (London: Faber and Faber)
3. Schwarz H 1966 *Pantheon* May/June
4. Seymour C 1964 *Art Bulletin* vol. 46
5. Fink, Daniel A II 1968 An attempt to determine a basis of J. Vermeer's method of painting by a comparison . . . *PhD thesis* Ohio University
6. Wheelock A K Jr 1978 *Perspective, Optics and Delft Artists Around 1650* (New York: Garland)
7. Coke, Van Deren (ed) 1975 *One Hundred Years of Photographic History* (Albuquerque: University of New Mexico Press)

8. Constable W G 1962 *Canaletto* (London: Oxford University Press)
9. Bindman D and Puppi L 1970 *Canaletto* (London: Weidenfeld and Nicolson)
10. Gioseffi D 1959 *Canaletto—e l'Impiego della Camera Ottica* (Trieste)
11. Carter B A R 1962 Review *Burlington Magazine* September
12. Links J G 1977 *Canaletto and his Patrons*
13. Tate Gallery 1973 *Landscape in Britain* (London)
14. Wallis M 1954 *Canaletto, the Painter of Warsaw* (Cracow)
15. Links J G 1972 *Townscape Painting* (London: Batsford)
16. Scharf A 1968 *Art and Photography* (Harmondsworth: Pelican)
17. Burke J 1955 *Wm. Hogarth, the 'Analysis of Beauty' with Rejected Passages* (London: Oxford University Press)
18. Whitley W T 1928 *Artists and their Friends in England 1700–1799* (London: Medici)
19. Hardie, Martin 1966, 1969 *Water Colour Painting in Britain* (London: Batsford)
20. Hilles F W 1929 *Letters of Sir Joshua Reynolds* (London: Cambridge University Press)
21. Oppe A P 1947 *Drawings of Paul and Thomas Sandby* (London: Phaidon)
22. Latimer J 1887 *Annals of Bristol in the 19th Century* (Bristol)
23. Hardie M and Clayton M 1932 *Walker's Quarterly* Nos. 35 and 36
24. Gernsheim H and Gernsheim A 1956 *L.J.M. Daguerre, A History of the Diorama and Daguerrotype*
25. Klingender F D 1947 *Art and the Industrial Revolution* (Cannington)
26. Cheselden W 1773 *Osteographia* (London)
27. Gernsheim H 1969 *History of Photography* (London: Oxford University Press)
28. Bruce J 1790 *Travels to Discover the Source of the Nile* (Edinburgh)
29. Playfair R L 1877 *Travels in the Footsteps of Bruce* (London)
30. Meteyard, Eliza 1866 *The Life of Josiah Wedgwood* (London: Hurst and Blackett)
31. Kelly, Alison 1975 *The Story of Wedgwood* (London: Faber and Faber)
32. Wills G 1969 *English Pottery and Porcelain* (Guinness Signatures)
33. Mason W 1819 *The English Garden* (London)
34. Rawson, Philip 1969 *Appreciation of the Arts* No. 3 (London: Oxford University Press)
35. White, John 1957 *Birth and Rebirth of Pictorial Space* (London: Faber and Faber)
36. Pirenne M H 1970 *Optics, Painting and Photography* (London: Cambridge University Press)
37. Clark K 1949 *Landscape into Art* (London: Murray)

38. Pollack P 1969 *Picture History of Photography* (New York: Abrams)
39. Eder K M 1945 *History of Photography* (New York: Columbia University Press)
40. Gombrich E H 1960 *Art and Illusion* (London)
41. Waterhouse J 1901 *Photographic Journal*

The Eighteenth Century

> Having gotten such a glass, make choice of some room which hath a north window.

In 1712 Joseph Addison wrote an essay for his daily paper *The Spectator* on the 'Pleasures of the Imagination'. He described a visit to a camera obscura in Greenwich Park where he found 'The prettiest landskip I ever saw was one drawn on the walls of a dark room.'[1] He referred to it as a common optical experiment and wrote: 'Here you might discover the waves and fluctuations of the water in strong and proper colours, with a picture of a ship entering at one end and sailing by degrees through the whole piece.' Since the picture was on a wall it was probably upside down, but Addison made no comment to this effect; neither did he mention the situation other than to say that it 'stood opposite on one side to a navigable river and on the other to a Park.'

Two years earlier, in 1710, Zacharias Conrad von Uffenbach and his brother were making a tour of England.[2] At this time two camera obscuras had been installed at the Royal Observatory at Greenwich and Uffenbach wrote: 'Leading out of the room is a small paved open vestibule or balcony, from which Mr. Flamsteed [the Astronomer Royal] makes his observations. On either side are two "camerae obscurae" which are uncommonly pleasant on account of the charming prospect and the great traffic on the Thames. We had an excellent view because it was fair, clear weather, so that there was no coal-smoke or fog in the air.' After their visit to the observatory they made their way to the Tower of London to see the jewels and historic documents, their guide being 'a man of some thirty years of age, who was, for an Englishman, most polite and discreet.' After conducting them round the Tower, the guide took them to his living quarters on an upper floor where he closed the shutter of one of the small rooms 'making of it a fine cameram obscuram.' Uffenbach relates how they again saw the boats and busy life of the Thames. They examined the lenses in the window shutter with considerable interest, and the servant said he had bought them from a Mr Marschall. They called on Mr Marschall who showed them that his name, painted on a rooftop some twenty houses away, could be easily read from the shop using a perspective glass or telescope. When recording this visit, Uffenbach said he would write further of his brother's poor opinion of English glass-cutting but, unfortunately, he did not do so in spite of his brother's many subsequent visits to see Mr Marschall. 'Glass-cutting' refers to the grinding and polishing of lenses at which his brother was an accomplished amateur.

It is more than likely that the Mr Marschall mentioned by Uffenbach was the instrument maker 'John Marshall, at the Sign of the Archimedes and two Golden Spectacles in Ludgate Street, near St. Paul's Churchyard, London, the House being new-built. Optician established 1690.'[3] So read his trade card, which also stated the approbation of the Royal Society. There were many instrument makers in London and their trade cards usually listed the apparatus they made. Telescopes, spectacles, microscopes and other optical instruments were prominent, although camera obscuras appear on some fifteen or so cards in the Science Museum collection. A footnote in Joseph Harris' *Treatise on Optics* (1775) commented that: 'These small cameras neatly and conveniently made, are now to be had at our shops.'

John Charnock (1764–1807), who lived at Blackheath not far from the Royal Observatory, had led a varied though insecure and somewhat eccentric life. He is perhaps best known for his *Biographia Navalis* which, although based mainly on hearsay, was nevertheless an excellent work on naval architecture and was well received at the time. Charnock later made many drawings of the machinery, apparatus and telescopes at the Royal Observatory which are now in the Royal Maritime Museum at Greenwich. Although most of the drawings are bound into books there are two separate sheets without any notes or identification which depict a camera obscura and its accessories such as a turret with a 45° mirror and a rotating mechanism, and also details of a method for making enlarged drawings of medallions.

The use of a camera obscura to make enlarged drawings of small objects was illustrated by Georg Friedrich Brander, who wrote a number of books on the instrument and its applications.[4] One of his engravings shows a statuette much closer to the lens than the camera obscura extension, and an enlarged image of the head is shown on the drawing surface. It was a massive reflex type of camera obscura which stood

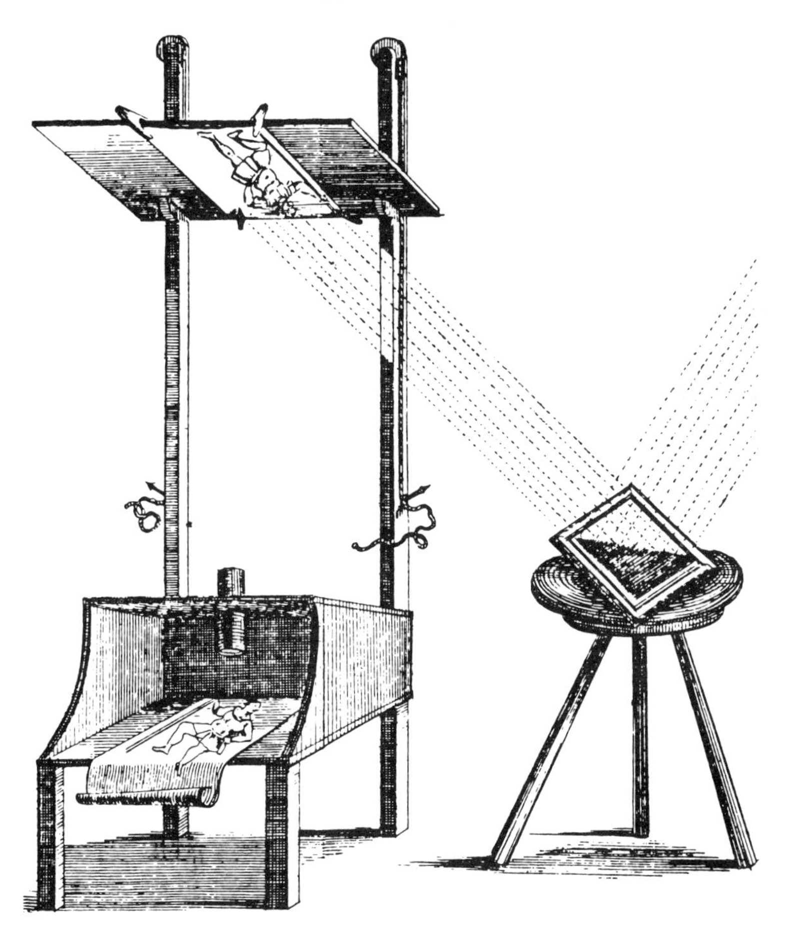

47. A camera obscura designed by Nicolai Bion in 1727 for copying drawings. The apparatus was used in a darkened room except for a shaft of sunlight which was reflected by a mirror onto the picture to be copied. Although the engraver has correctly depicted the reversal of the copy, he has mistakenly omitted to invert it.

48. A camera obscura from a book by G F Brander published in 1769. The very long extension of the lens and the close proximity of the statuette suggests that it illustrates a method for making an enlarged image. In this instance only the head of the figure would appear on the screen of the camera obscura.

about four feet high. Brander (1713–85) was a pupil of Doppelmeier, and he studied mathematics and physics at Nuremburg and Altdorf. From 1734 he lived at Augsburg inventing and making surgical, physical and astronomical instruments.

The large box-type camera obscura as used by Brander and also by Cheselden for his book *Osteographia*, was ideal for making drawings for the connoisseur or engraver. At this time, however, there was a growing demand for more easily portable instruments, since this was the century of the 'Grand Tour' and travelling for pleasure had become the vogue. Although it was none too easy to transport, a camera obscura was needed to record the views and sights during a tour, and some form of tent, as Kepler had constructed, was usually adopted.

Mr Dugal Campbell, an engineer in His Majesty's Service, made a camera obscura by attaching four poles, each about three feet long to the corners of a square board.[5,6] The opposite ends of the poles were then fixed to the corners of another, larger board, so that the whole construction

49. An early eighteenth-century camera obscura at the Museum of the History of Science, Oxford. The decorated lens tube is typical of the late seventeenth and early eighteenth centuries when microscope and telescope tubes were made of pasteboard and frequently covered with leather. Copyright—Museum of the History of Science, Oxford.

resembled a truncated pyramid. The smaller top board had a mirror at an angle of 45° reflecting an image through a hole containing a lens onto the baseboard. Dark leather or cloth was hung from the sides of the top board so that, by raising one of these curtains, the viewer could insert his head and hands to trace the image.

Mr Thompson of Duddingston made a portable camera obscura which could be folded up like an umbrella.[5] This was probably similar to the camera obscura made by Abbe Nollet which was submitted to the French Royal Academy of Sciences and awarded a patent.[7] Nollet's apparatus was a very small tent consisting of four rods converging from the corners of a base-board to form a skeleton pyramid. A mirror and a lens at the apex of the pyramid produced an image on a sheet of paper placed on the base-board. A cloth with a hole in the centre was draped over the framework, and a drawstring at the hole was tied at the lens mounting. The tent could be mounted on a table in a bay window where one could sit comfortably to trace the projected view from the window. A similar small tent and a sedan chair camera obscura are illustrated in Middleton's *New Complete Dictionary of Arts and Sciences.*[8]

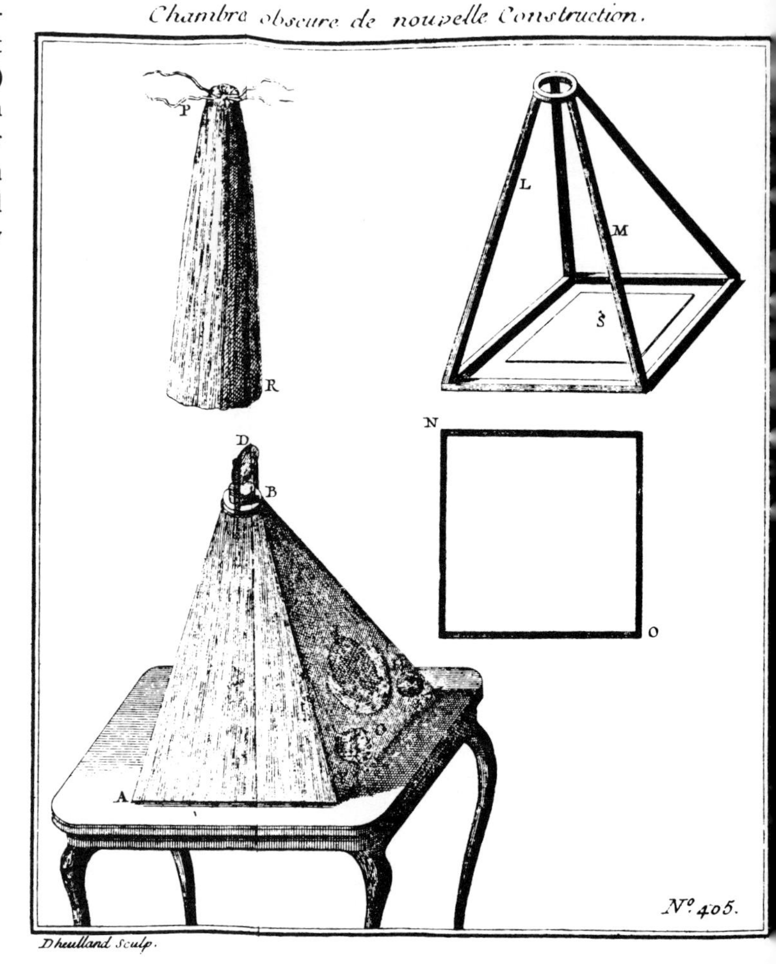

50. Abbe Nollet's design for a collapsible camera obscura (1752). That in the engraving by Costa (p49) appears to be of this type. Joseph Harris wrote in 1775 that a Mr Thompson of Duddingston had made a camera obscura which could be folded up like an umbrella; it was probably similar to this design.

Instrument Makers

A famous instrument maker of the period, John Cuff†, whose premises were situated next to Sergeant's Inn in Fleet Street, advertised on his trade card, together with many other instruments, 'camera obscuras for exhibiting prospects in their natural proportions and colours.'[9] Unfortunately very little is now known about his camera obscuras, but the British Museum has a booklet entitled

†A portrait of John Cuff by Zoffany now hangs in Windsor Castle.

51. A room camera obscura made in 1754. A similar engraving, different only in the shape of the trees and other minor details, also appeared in Middleton's *New Complete Dictionary of Arts and Sciences* (1778).

'Verses occasioned by the sight of a chamera obscura, printed for John Cuff, 1747.' There is no author's name and the verse is in a simple style, relating the history of Bacon's trial for witchcraft and describing the brilliant colours of the garden with crystal fountains, all shown on the 'clear white sheet.' The poem continues with a story of a gallant fleet that rode the swelling tide but, 'Ah! trust no summer's sea, nor Harlot's smile.' The verse describes a stormy sea and the destruction of the fleet, but the picture faded when the door was opened and the sun's bright rays shone in. These stanzas suggest that the camera obscura showed enacted scenes similar to those decribed by Porta. Early in the poem there are lines which make one wish that John Donne, the metaphysical poet and divine, had seen a camera obscura:

> How the Clown stares! Smit with surprise and love,
> To see th'inverted pretty Milk-Maid move,
> With Pail beneath her head, and feet above.

John Dollond (1706–61), whose name is still well known, was an optician who had a workshop near Exeter Exchange in the Strand, London.[10] Dollond made many improvements to lens design and was the first to construct an achromatic lens in about 1758 which minimised colour fringes and produced a clearer image in the camera obscura.

JOHN CUFF,

Optician, Spectacle, *and* Microſcope *Maker*,

At the Sign of the REFLECTING MICROSCOPE and SPECTACLES, againſt *Serjeant's-Inn* Gate in *Fleet-ſtreet*,

MAKES and Sells all Sorts of the moſt curious Optical Inſtruments, ſuch as SPECTACLES and READING GLASSES, ground on Braſs Tools in the Method approved by the Royal Society, of *Brazil* Pebbles, Rock Chryſtal, or the fineſt Flint Glaſs, and ſet either in Silver, Mother of Pearl, Tortoiſe-ſhell, *&c*, in the moſt convenient Manner for Uſe.

REFRACTING TELESCOPES, proper for celeſtial or terreſtrial Obſervations, either by Land or Sea; and alſo REFLECTING TELESCOPES, as invented by Sir *Iſaac Newton* and Mr *Gregory*.

MICROSCOPES of ſeveral Kinds, both Single and Double, (either for the Study or the Pocket) particularly a new-conſtructed DOUBLE MICROSCOPE, invented by the ſaid *John Cuff*, to remedy the Inconveniencies complained of in all the former Sorts; the SOLAR MICROSCOPE, as improv'd by him with a moſt convenient and manageable Apparatus for viewing Objects, and a Contrivance for drawing the Pictures of them with great Eaſe and Exactneſs, even by thoſe that have never drawn before; the AQUATIC MICROSCOPE, invented alſo by him for the Examination of Water Animals; the OPAKE MICROSCOPE, which by a Light reflected upon Objects that have no Tranſparency, renders them diſtinctly viſible; *Culpepper*'s MICROSCOPE, *Wilſon*'s Pocket MICROSCOPE, *&c*.

The CAMERA OBSCURA for exhibiting Proſpects in their natural Proportions and Colours, together with the Motions of living Subjects: MAGIC LANTHORNS; Convex and Concave SPECULUMS or LOOKING GLASSES, MULTIPLYING GLASSES, BAROMETERS, THERMOMETERS, OPERA GLASSES, SPECTACLES of true *Venetian* Green Glaſs, SPEAKING TRUMPETS, ſeveral Mathematical Inſtruments, and many other Curioſities, are ſold by the ſaid Operator at reaſonable Prices, either Wholeſale or Retale.

52. The trade card of John Cuff who published a book of verses about the camera obscura.

His workshop produced a great variety of surveying instruments and telescopes for the army and navy, and although very little is now known about them, he also made camera obscuras.

William Storer, another London instrument maker, obtained a patent for a reflex camera obscura of advanced design in 1778, which he called the 'Royal Accurate Delineator', patent number 1183.[11] The Science Museum has one of these instruments and a copy of the book describing it, entitled 'Storer's syllabus, to a course of optical experiments, on the syllepsio optica; or the new optical principles of the Royal Delineator Analysed'. This camera obscura was constructed of two boxes, one sliding within the other by a rack and pinion movement. The lens was compound, consisting of two large rectangular biconvex lenses to produce a bright image which was further enhanced by a third lens below the ground-glass screen. This lens increased the illumination at the corners and edges of the screen in the same way as the Fresnel lens makes a bright viewfinder image in a modern photographic reflex camera. Although the image in Storer's camera obscura was exceptionally clear, the large-aperture lens permitted only a shallow depth of sharp picture. The patent for the Royal

53. William Storer's camera obscura, the 'Royal Accurate Delineator' (1778). This instrument had an optical system superior to any camera obscura of the period. The image was exceptionally bright, which made it suitable for sketching portraits.

Accurate Delineator was printed in the *Mechanics Magazine* (1845) and Storer's address was shown as Lisle-street, Leicester-fields, Middlesex.

Horace Walpole (1717–97), novelist and designer of Strawberry Hill, his 'little Gothic castle' at Twickenham, had many friends who used or were familiar with a camera obscura, and he bought a delineator from Mr Storer.[12] In a letter to the Reverend William West he wrote:

> This is but a codicil to my last, but I forgot to mention in it a new discovery that charms me more than Harlequin did at ten years old, and will bring all paradise before your eyes more perfectly than you can paint it to the good women of your parish. It will be the delight of your solitude, and will rival your own celestinette. It is such a perfecting of the camera obscura, that it no longer depends on the sun, and serves for taking portraits with a force and exactness incredible; and serves almost as well by candlelight as by day. It is called the delineator, and is invented within these eighteen months by a Mr. Storer, a Norfolk man, one of the modestest and humblest of beings. Sir Joshua Reynolds and West are gone mad with it, and it will be their own faults if they do not excel Rubens in light and shade, and all the Flemish masters in truth. It improves the beauty of trees,—I don't know what it does not do—everything for me, for I can have every inside of every room here drawn minutely in the size of this page. Mr. Storer fell as much in love with Strawberry Hill as I did with his instrument. The perspectives of the house, which I studied so much, are miraculous in this camera. The Gallery, Cabinet, Round Drawing Room, and Great Bed Chamber, make such pictures as you never saw. The painted glass and trees that shade it are Arabian tales. This instrument will enable engravers to copy pictures with the utmost precision: and with it you may take a vase or the pattern of a china jar in a moment; architecture and trees are its greatest beauty; but I think it will perform more wonders than electricity, and yet it is so simple as to be contained in a trunk, that you may carry in your lap in your chaise, for there is such contrivance in that trunk that the filberd in the fairy tales which held such treasures was a fool to it. In short it is terrible to be three score when it is just invented; I could play with it for forty years; when will you come up and see it? I am sure you will not go back without one.

The letter was dated Strawberry Hill, 21 September 1777 when Walpole had yet twenty years to play with his camera obscura.

Benjamin Martin's workshop was situated near Crane Court in Fleet Street; besides making scientific instruments he wrote and published a number of works which included six volumes of *Biographia Britannica* and a *General Magazine of Arts and Sciences,* which ran to fourteen volumes published between 1755 and 1765.[13,14] In his book *Biographia Philosophica* he makes a terse observation on Porta, 'His Natural Magic, a book abounding in curious experments; but contains nothing of Magic, in the common acceptation of the word, as he pretends to nothing above the Power of Nature.' Martin also wrote a popular book of scientific instruction in the question/answer style. Euphrosyne is the student and her teacher is Cleonicus; both have a good deal to say for themselves. Cleonicus discourses at length on design and drawing, to which Euphrosyne replies: 'This have I oftentimes, with the highest pleasure, observed, and I think nothing can compare with the beauty and perfection of the landscapes that I have seen formed by large convex speculums. Those that you showed me in the darkened room, some time ago, are of the same kind, but inverted. In short, I often amuse myself with the picturesque appearance of objects, by reflection of almost every kind of glass that comes my way'. Her large convex speculum was a Claude Lorraine glass.

In a more serious book entitled *A New and Compendious System of Optics* (1740), Martin devoted a chapter to the camera obscura, the scioptric ball and methods for correcting the inverted image, and also referred to the best aspect for a room to be used for a camera obscura.[15] About the image made by a camera obscura he noted, 'This is Nature's Art of Painting, and it is with ease observed, how infinitely superior this is to the finest performance of the pencil.' He continued by describing the appearance of birds, flowers, carts, ships and many other objects in the camera obscura: 'These are the inimitable perfections of a picture drawn by nature's hand; in comparison of which, how mean, how coarse, how imperfect, yea, what sorry daubing is the finest artificial painting.' Later in the same chapter he remarked that the camera obscura was useful to the optician for demonstrating 'diverging and converging rays in any degree, and so can prove the truth of all those properties of convex, concave, and meniscus glasses, with respect to these rays, as the theory teaches.' Finally, Martin explained how to make a rainbow: 'In short, this is a good experiment towards explaining the nature of the celestial bow.'

The 'Artist' camera obscura made by Jones has already been mentioned, but a similar instrument bearing the same

54. The Harvard University book-form camera obscura supplied by Benjamin Martin after a fire at the university in 1764. When folded to book shape the size was about 24×18×5 inches. Martin invoiced the instrument at £3 13s 6d.

inscription was described in a letter with an accompanying illustration in the *British Journal of Photography* of 1 December 1916.[16] The issue for the following week contained an editorial note about the letter: 'The "Jones, Artist" whose name is stamped on the camera is probably one of the two Jones' of the firm, 'next Furnival's Inn' from whose price list we quoted. The date of this list (1797) is also that of the camera, so the latter was probably one of the latest patterns at that time.' The price list is that of W & S Jones, and is an interesting example of the range of camera obscuras which were then available. Pocket instruments varied in price from 9s to £1 16s 0d, and a large one which folded up to look like a book cost £8 18s 6d. A newly invented, 'very portable' camera obscura with sides of canvas which allowed it to be folded, was priced at £2 2s 0d. Prices were also printed on the trade card of 'Henry Pyefinch at the Golden Quadrant, Sun and Spectacles, No. 67, between Bishopsgate Street and the Royal Exchange in Cornhill, London. A pocket camera, 10s 6d, ditto larger 15s 0d. Camera obscura in the form of a book, the best £4 4s 0d. The same adapted to view prints £5 5s 0d'.[3] By equating the cheapest camera obscura with the cheapest photographic camera of today, the price range was equivalent to approximately £10 to £160.

Robert Smith in his *Compleat System of Optics* (1738) included a chapter on 'Optical machines for making pictures of objects; and their uses in drawing.'[17] After dealing with the camera obscura he explained a 'sky-optric' ball and the 'broadest lens of this kind that I ever saw, is at Mr. Scarlett's shop by St. Anne's Church.' Scarlett demonstrated the lens

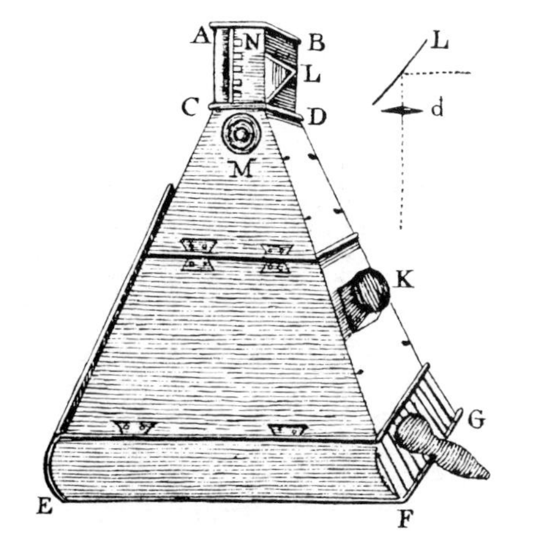

55. A book-form camera obscura from the *Encyclopaedia Britannica* (1794). The whole apparatus, when collapsed into the base EFG, looked like a very large book. The drawing at the side illustrates the mirror L and the lens d in the box ABCD. The unit moved up and down by means of the knob M in order to focus the image, and this was viewed through the hole K. The hand could be inserted through the sleeve FG to trace the image. Book-form camera obscuras were priced from £5 to £10.

by projecting the street scene into his shop which, in the days of small windows, was probably dark enough for the image to be seen clearly. Robert Smith was diverted and surprised at the inverted picture which showed the perpetual dancing and undulating motion of persons walking. He was very curious about the bobbing movement and reasoned that it really did happen: '. . . for I question whether he could walk freely between two parallel boards, if the upper board be but a little higher than his head when he stood still . . .' Finally he inclined a mirror to the screen in order to correct the image and 'the people in the street will

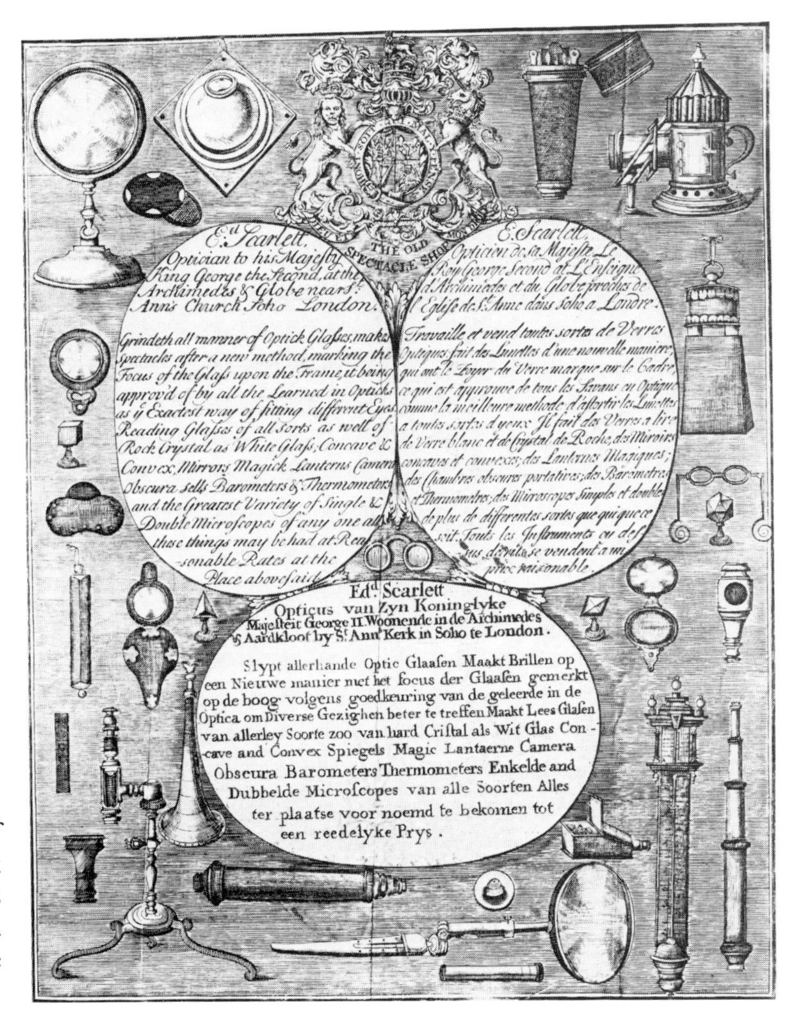

56. Edward Scarlett's trade card of about 1758 which advertised his work in three languages. A scioptric ball is shown above the English panel, and a camera obscura at the lower right of the Dutch panel.

appear upright and without any undulating motion of the heads.' The chapter continued with descriptions of two camera obscuras, one of which was 'made specially by Mr. Scarlett for the business of drawing.' Edward Scarlett's trade card stated that he was at the 'Archimedes and Globe near St. Anne's Church, Soho, London, Optician to his Majesty King George the Second.'[3] The card was printed in English, French and Dutch, each language in an oval panel surrounded by engravings of the instruments available from his workshop.

It is possible that Scarlett's camera obscura display in his shop window may have been imitated by other shopkeepers. A camera obscura in a window was mentioned by William Weller Pepys, Master in Chancery, in a letter to Mrs Montague the famous bluestocking†. His letter was addressed from Wimpole Street, 22 July 1782.[18] 'I hope you have read Cecilia, and that you are as much pleased as I am with the delineation of some of the characters; Mrs. Thrale told me the other day that it seemed to her like a camera obscura in a window in Piccadilly, so exactly were the characters represented.' Mrs Thrale was a friend of Fanny Burney, the author of *Cecilia,* and was well known in the literary circles of the day.

Ferguson's series of *Lectures on Select Subjects* was published in 1764 and continued through many editions until 1805, it then came under the editorship of David Brewster and further printings were made.[19] The entry for the camera obscura remained unchanged from 1764 to 1825. A diagram showing how the image was made was carefully annotated and explained in the text, followed by a description of a method for magnifying the image by placing a convex lens after the mirror. 'In this manner, the opera-glasses are constructed; with which a gentleman may look at a lady at a distance in the company, and the lady know nothing of it.'

†Blunt R 1923 *Mrs. Montague, Queen of the Blues.*

Early in the eighteenth century W J s'Gravesande wrote a book on perspective and explained how to use a camera obscura for drawing.[4] One of the illustrations showed a sedan chair converted to a camera obscura. Jombert, writing some years later, quoted Gravesande but warned of the distortion of colours in the camera obscura, and that nature should be painted as seen normally and not with the heightened effect produced by the camera obscura.

Dr Charles Hutton translated the book *Recreations in Mathematics and Natural Philosophy* (1803) in which he provided dimensions for constructing the sedan-chair camera obscura of Gravesande.[20] He included a light-proof tube to provide fresh air and explained how to obtain successive views around the camera obscura by revolving the mirror. The Royal Observatory at Greenwich was referred to as having an excellent camera obscura allowing five or six persons to view the picture together. 'The images are thrown on a large smooth concave table cast of plaster of Paris, and moveable up and down so as to suit the distance of the objects.' Hutton also compiled a mathematical dictionary in which he explained the theory of the camera obscura. He again referred to the Royal Observatory and that the camera obscura had a lens and mirror in a turret which could be moved round by rods in order to provide successive views of the horizon. He also referred to the book *Optics* by Harris, and to Storer's patent camera obscura, the 'Delineator'.

The article on the camera obscura in Chambers *Cyclopaedia* (1741) stated that the invention of the instrument could be ascribed to Baptista Porta, and continued with details of the construction and uses of a room camera obscura and two portable instruments. Mention was also made of the usual preparation of the room by covering the windows and making a hole for the lens, and noted that 'if the hole be about the size of a pea, an image will be made without a lens.' This statement was repeated almost word for word in Peter Barlow's dictionary published in 1814. The text and illustrations in Chambers *Cyclopaedia* are very clear and would not have presented any difficulty to a craftsman at the time.

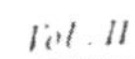

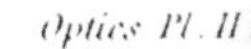

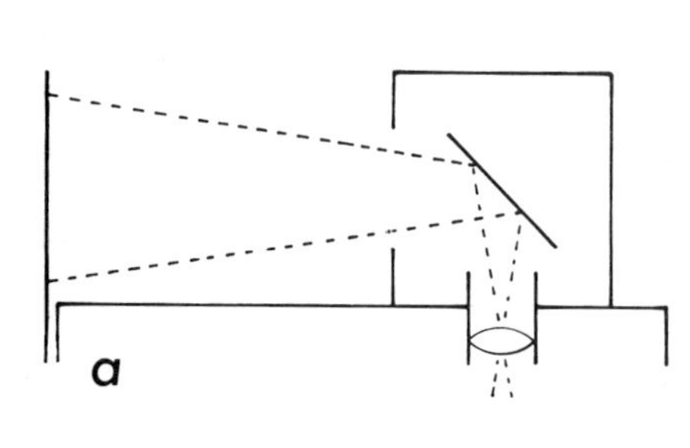

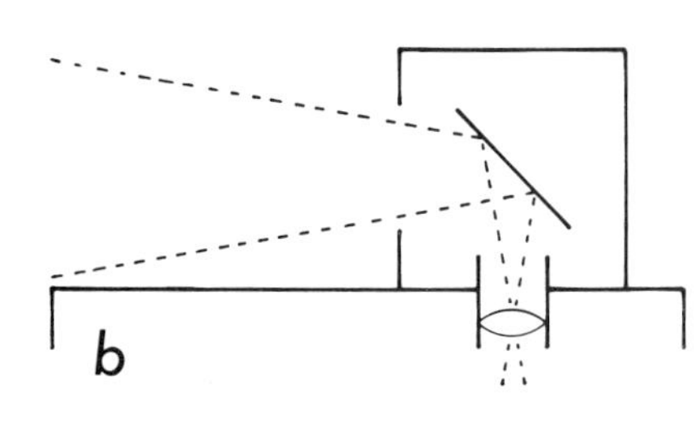

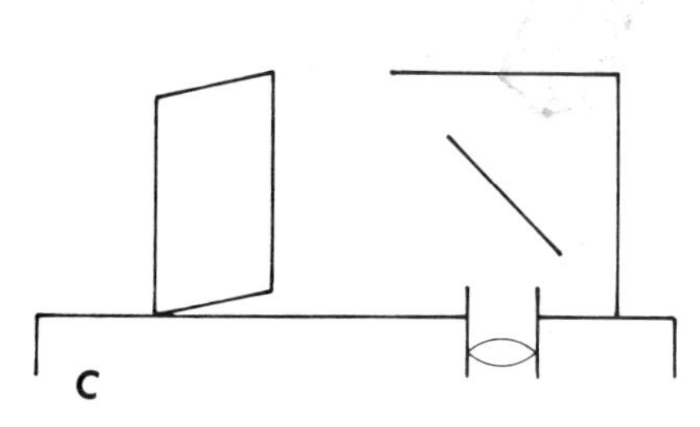

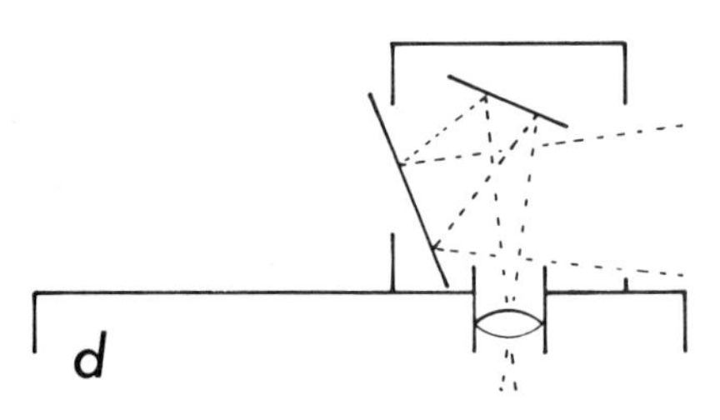

57. Illustration of an all-purpose sedan chair camera obscura which appeared in *An Essay on Perspective* by W J s'Gravesande (1711). In 1814 Dr Hutton described the construction and use of this instrument: the mirror H rotated on a vertical axis, and an easel D supported the prints or paintings to be copied, when the mirror would be removed (a). When both H and D were removed a view to the rear was projected onto the table, where it appeared the right way up and correct from left to right (b). A panorama was made by replacing the mirror H and rotating it for each view (c). By placing the mirror H in the box X and adjusting the other mirror, an image of the forward view would appear upright but reversed from left to right. Fig. 8 in the engraving is an air tube; Hutton commented that further ventilation could be provided by a foot-operated bellows.

In his *Complete Dictionary of Arts and Sciences,* Croker enumerated ten requirements for producing a good picture in a camera obscura. He referred to the lens as being in a scioptric ball and that it should be free from veins and blebs. He pointed out that the image should be received on an 'elliptical' surface 'but this is a nicety which few will think worth regarding, who do not aim at a very great accuracy indeed in what they do.' Croker also mentioned that if an object is placed at twice the focal length from a lens, the image will be made at the same distance and will be the same size.

A M Guyot in his book *Nouvelles Récréations Physiques et Mathématiques* (1770) illustrated a camera obscura constructed as an inverted pyramid requiring four legs for support.[21] The apex of the pyramid was very near the floor and contained the lens and mirror. This instrument is frequently illustrated in dictionaries of photography and is usually described as a table or desk camera obscura. It is often referred to as 'curious' because it appears to have a very low viewpoint. The dimensions given by Guyot indicate that it was only about 24 inches high, so it was probably intended for use on a low table or stool and indoors at a window. A gap in the drawn curtains would allow the lens to take a view through the window, and the darkened room would enhance the image brightness. The drawing surface was quite large, 24 × 24 inches. Dr W Hooper, in his book *Rational Recreations,* provided instructions on how to make Guyot's camera obscura, noting that it had some advantages not to be found in others.[22] In respect of our lack

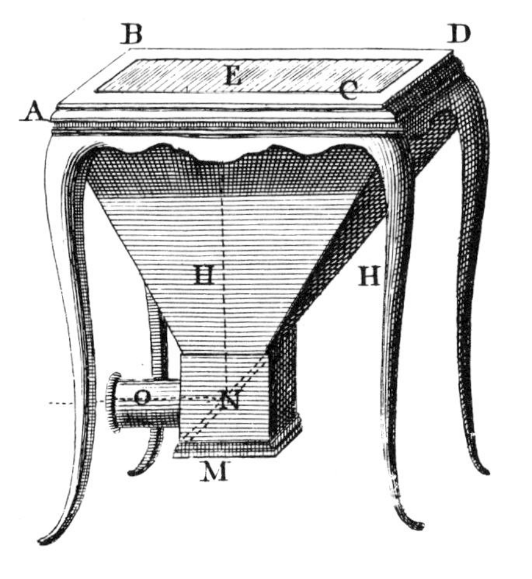

58. A camera obscura designed by M Guyot in 1770, described in detail by Hooper in *Rational Recreations* (1787). The top frame measured 20×24 inches, the legs hinged inwards for carrying, and four poles placed at the corners of the frame draped with a black cloth made a small tent which could be opened at one side for viewing the image. Hooper recommended that the instrument be placed on an elevated spot, 'that nothing may intercept the rays falling on the glass lens'. The advantage of this design was that 'you may draw without having your hand between the rays and their image'.

of information about the use of a camera obscura for book illustration, Hooper's final note is of interest:

> By each of these methods you will have the objects either in their natural position, or reversed which will be an advantage when the design is to be engraved, and you will have it then appear in the natural position.

The reference to engraving a picture can only mean that a large number of copies would be printed for supply to picture and print shops or to book publishers.

Early in the eighteenth century John Harris compiled the first alphabetical encyclopaedia, *Lexicon Technicum,* in which he was most enthusiastic about the camera obscura.[23] He noted that glasses (lenses) which drew at 15, 20 or 25 feet were not suitable; besides, '. . . such large glasses likewise are not easily had everywhere, nor are they every one's money; but a glass that draws about 6 feet is very proper to be used in this case. Having gotten such a glass, make choice of some room which hath a north window . . .' He then went on to explain how to black out the room, to fix the lens and to hang up a white sheet and 'move it to and fro till you find the objects are represented on it very distinctly, and then you may fasten it there by nails to the ceiling.' Harris was full of admiration for the picture: 'Another thing in which this representation exceeds painting is that you have motion expressed on your cloth.' He then recounted his joy at seeing the images of flowers and plants move in the wind, flies and birds in the sky and persons walking: 'In a word nothing is wanting to render it one of the finest sights in the world, but all things are inverted, and the wrong end upwards.' This could be corrected, he explained, by using two lenses, but 'not so easy as to take a common looking glass of about 12 or 14 inches square and hold it under or near the chin with an accute angle to your breast.' This arrangement 'gives it a glariness that is very surprising and makes it look like some magical prospect and the moving images like so many spectrums or phantoms. And no doubt there are many persons who would believe it to be no less than downright conjuration.' The following paragraph, about cunning men and cheating knaves, is reminiscent of the earlier exploits of Cardano and Drebbel. Harris also mentioned a lens in a globe of wood (the scioptric ball), 'which like the eye of an animal may be turned everyway' and that it could be obtained from Mr Marshall at the Archimedes on Ludgate Hill. Harris concluded by explaining Hooke's method of projecting an image of an object into a room which did not need to be darkened. The object should be inverted, but 'If the object cannot be inverted (as 'tis pretty difficult with living animals, candles etc.) then there must be two large glasses . . .' This experiment of Robert Hooke's is more properly part of the history of the magic lantern†.

Another Harris, Joseph, Assay Master at the Royal Mint, wrote his *Treatise of Optics* in 1775 explaining, with excellent diagrams, the formation of an image by a lens.[6] He also mentioned ground glass (rough on one side), the scioptric ball (by some called an ox-eye), and he gave instructions on how to make a camera obscura. Although much of this article simply repeated many other previous descriptions, Harris did include some novel ideas of his own. He suggested making a screen on a roller and looking at the picture on the screen through a small opening from an adjacent room which need not be darkened. He described how to make a camera obscura for drawing, and a show box perhaps with a base to take several hundred prints or a book, which could be easily converted to a camera obscura. For persons able to draw freehand, Harris believed that a pocket camera obscura would suffice all their purposes. 'These small cameras neatly and conveniently made are now to be had at our shops. They have been made with a lens of about 1½ or 2 inches focus, and fitted to the heads of canes; but these are much too small to answer any purpose.' No doubt a camera obscura so small as to be fitted into a walking stick would have been too small for any purpose except, perhaps, for unobtrusively looking at a lady in the company!

†cf. *Philosophical Transactions of the Royal Society* No. 38 p741, 17 August 1668.

59. A small camera obscura made by the furniture maker Thomas Sheraton (1751–1806). This was probably a special commission for a gift; with its inlaid banding and marquetry, it was probably no more than an elegant toy, measuring approximately 2×3×3 inches.

A small camera obscura 2 × 3 × 3 inches in size was illustrated by Gross and is attributed to the furniture maker and designer Thomas Sheraton; unfortunately a large volume of Sheraton's designs has given no further evidence of camera obscuras.[24] It was probably a single commission and, from its appearance, was a likely present for a lady. The lens was fixed at the front of the box and at the rear was an inner box which could be withdrawn, containing the mirror and ground-glass screen. The movement of one box within the other provided the necessary extension for focusing the lens. All the edges of the camera obscura have inlaid banding and the top has an oval of marquetry. It is a charming little instrument.

Joseph Harris referred to several people concerned with camera obscuras: Mr Dugal Campbell, Mr Parrat of Killerton near Kirby Lonsdale, Mr Johnson of Berwick (who had a show box at Charing Cross), Mr Nairn, instrument maker in Cornhill, Mr Short, who had much experience in the making of mirrors, and Dr Hoadley at Chelsea.[6]

Dr Benjamin Hoadley was born on 20 February 1705, the eldest son of the Bishop of Winchester. He went to Cambridge, became a Fellow of the Royal Society when quite young and was appointed physician to His Majesty's household in 1742. In 1747 he built a house in Chelsea next to Chelsea Farm on the west of Lord Cremorn's estate, which then became known as Hoadley House, but when he died it was bought by Earl Ashburnham whereupon it became Ashburnham House. All that remains today is the street name—Camera Place, SW10. During the last century, however, there was Camera Square, Camera Street, Little Camera Place and Camera Terrace. The camera obscura must have been a very well known feature of Chelsea but unfortunately we have only Joseph Harris' description:

> A complete thing of this kind is a little room built by Dr. Hoadley of Chelsea. The roof is pyramidical like that of a summer house, and out of the sides of it are projections like those of dormer lights or windows, in each front is placed a lens

> with a plane metallic speculum [mirror] fixed at the proper angle for throwing the picture upon a large glass half polished [ground glass], fixed against the cavity made by the said projection close to the ceiling. There is another apparatus of the same kind at the top, the lens and little speculum being made to turn all round, so that there are five pictures in the room at once, and which give sufficient light for reading small print.

Although Hoadley was a Fellow of the Royal Society and had published papers in the *Philosophical Transactions,* he apparently wrote nothing about his camera obscura. The use of ground glass as a translucent screen and of four pictures around the room was merely an extension of Kircher's double camera obscura of the seventeenth century, and was a forerunner of back-projection and multi-screen presentations of today. The fifth picture from the lens in the apex of the roof may have been projected onto a table, but there seems every reason to suppose that Hoadley again used ground glass and made a ceiling picture. A note in the *British Journal of Photography* of 8 December 1916 suggests that Hoadley's camera obscura at Chelsea may have been used by the general public.

Joseph Harris probably knew of Mr S Parrat of Killerton, near Kirby Lonsdale, through a letter which appeared in the *Gentleman's Magazine* of April 1753.[25] Parrat started his letter in a manner which is no less familiar today, 'As I have for some years been a reader of your magazine, and always have a pleasure to behold every attempt to increase our knowledge . . .', and went on to describe in great detail a show box he had made for viewing prints, engravings, etc. This instrument had the advantage of being easily converted to a camera obscura by which many objects could be 'represented on the paper with such exquisite exactness as far surpasses the utmost skill of any painter to express.'

John Hinton published the *Universal Magazine* monthly from the King's Arms in Newgate Street, London.[26] The issue for May 1752 contained an article describing the camera obscura and its uses for drawing. The article was accompanied by a full-page copper-plate engraving showing

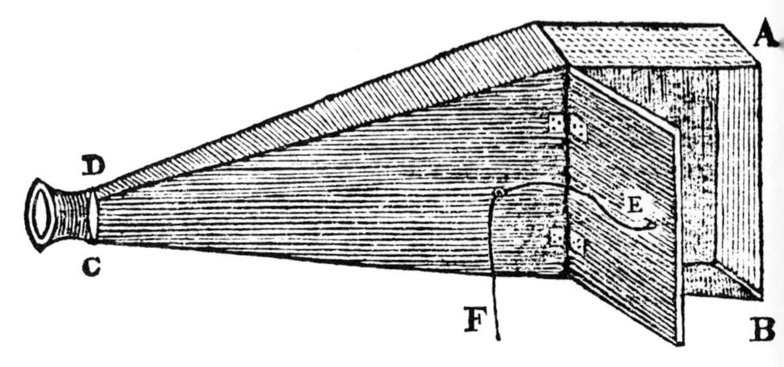

60. Parrat's camera obscura, described in a letter to the *Gentleman's Magazine* in April 1753. The lens CD had a focal length of about 30 inches and was 4 inches in diameter. The base AB, about 18 × 10 inches, was covered with a white paper screen and the image was viewed by lifting the flap E. When the apparatus was used as a show box the underside of the flap was covered with white paper to reflect light onto a print or picture placed on the base. The string F was tied to hold the flap open at a suitable angle for reflection. The picture was viewed through the lens 'with both eyes and you will see the picture, at a vast distance, surprisingly magnified.'

a cross section of a darkened room with an image of a garden and a man sitting near to an enormous urn. Hinton said that the camera obscura was of great use in explaining the nature of vision, and that experiments made with it were 'beautiful and diverting'. He continued with the usual description of the construction of a camera obscura and an explanation of how the image is made. The greater part of the article was taken up with a comparison and discussion of the three representations of a scene: a painting, a camera obscura and a looking glass. The last item is extended to the viewing of a painting in a looking glass. Hinton seemed to be aware of a correct viewing distance for a painting and that when observing objects in a mirror, 'they are more natural and just than those in a camera obscura. We shall only add, that the reflection of an object, in a looking glass at any distance of the eye, being the same with stereographical appearance of the object, to an eye placed at an equal distance perpendicularly on the contrary side of the glass, supposing it transparent.' Today we would probably say that an image in a mirror has the same appearance as when we look at an object from a position behind the mirror which is twice the distance between the mirror and the object.

In 1772 Joseph Priestley, scientist, non-conformist minister and reformer, wrote a *History and Present State of Discoveries Relating to Vision, Light and Colours*, in which he referred to Battista Porta as the inventor of the camera obscura.[27] After describing the theatricals and Porta's method of using drawings on thin paper illuminated from behind by the sun, in order to make pictures in a camera obscura, Priestley commented that this 'gave the hint to Kircher, on which he constructed the magic lantern, doing that in the night, and in many respects more conveniently, which Porta exhibited in the day.'

61. This illustration of a room camera obscura was printed in the *Universal Magazine*, May 1752. The title page said: 'Be careful to ask for the Universal Magazine, printed for John Hinton, at the King's Arms in Newgate Street, London.'

Writers

Alexander Pope, also known as 'Little Pope', the greatest of the Augustan poets, suffered from ill health throughout his life. In a letter written in 1725 he made a somewhat sad comment on his infirmities, 'for my whole life has been but one long disease,' but fortunately his literary success allowed him to live in reasonable comfort and he was able to afford the attention of two famous doctors.[28]

> I'll do what Mead and Cheselden advise,
> to keep these limbs,
> and preserve these eyes.

The house he built on the bank of the Thames at Twickenham was separated from the garden by a main road, so in order to get from one to the other he excavated a tunnel under the road. The tunnel provided a view of the Thames from an ornamental temple in the garden, and Pope turned the tunnel into a grotto with doors at each end. In a letter to Edward Blount he wrote: 'I have put my last hand to my works of this kind, in happily finishing the subterraneous way and grotto. I there found a spring of the clearest water, which falls in a perpetual rill, that echoes throughout the cavern day and night. When you shut the doors of this grotto it becomes on the instant, from a luminous room, a camera obscura, on the walls of which all the objects of the river, hills, woods, and boats, are forming a moving picture in their visible radiations; and when you have a mind to light it up, it affords you a very different scene.' It is almost certain that the picture in the grotto was made by a small hole or gap in the doors facing the sunlit northern bank of the river. Pope did not mention using a lens for the camera obscura in the grotto, but we do know that he was interested in scientific matters, especially astronomy. Dr Cheselden attended him medically and they met in the same literary circles with Sir Joshua Reynolds, Charles Jervais (another portrait painter) and Joseph Addison, all of whom were familiar with the camera obscura. John Gay, the poet and dramatist, was befriended by Pope and frequently attended the same literary meetings.

In a long heroic poem, *The Fan,* Gay referred to a camera obscura.

> Flocks graze the plains, birds wing the silent air
> In darkened rooms, where light can only pass
> Through the small circle of a convex glass;
> On the white sheet the moving figures rise,
> The forest waves, clouds float along the skies.

The poem occupies three books and Dr Johnson thought it was dull: 'one of those mythological fictions which antiquity delivers ready to the hand.'

A camera obscura was also mentioned by Samuel Richardson in *Sir Charles Grandison*[29] and by Laurence Sterne in *Tristram Shandy.*[30] Sir Charles Grandison and his sister were discussing her suitors: 'The subjects we are upon, courtship and marriage, cannot, I find, be talked of seriously by a Lady, before company. Shall I retire with you to solitude? Make a Lover's Camera Obscura for you? Or, could I place you upon the mossy bank of a purling stream . . .' In *Tristram Shandy* Laurence Sterne made verbal play of the phrase 'drawing of a character' and deviously entangles the artist. 'One of these you will see drawing a full-length character against the light;—that's illiberal,—dishonest,—and hard upon the character of the man who sits. Others, to mend the matter, will make a drawing of you in the camera;—that is most unfair of all,—because, there you are sure to be represented in some of your most ridiculous attitudes.'

Martin Froben Ledermuller of Nuremburg was an Officer of Justice and Secretary to the Imperial Forestry and Beekeeping Commission, and was also interested in microscopy.[31] His book *Microscopic Delights of the Mind and Eye* was illustrated with coloured engravings showing three uses for a solar microscope, one of which was the projection of magnified images of small insect specimens into a box-type camera obscura. The complete apparatus of the camera obscura and solar microscope was shown attached to a black area which appears to be a board in a window frame with a pelmet draped with coloured curtains. Simple and reflex camera obscuras are illustrated, each with a disembodied arm and hand holding a pencil on the drawing area. These engravings, though somewhat naïve, demonstrate an early attempt to produce accurate drawings of microscopic objects with the aid of a camera obscura. A similar technique was described in the nineteenth century.

The Victoria and Albert Museum in London has a book-form camera obscura which is said to be of French

62. A reflex (*top*) and a simple (*bottom*) camera obscura for tracing images projected by a solar microscope. (From Ledermuller's book on microscopy, 1760.)

manufacture dated about 1750, probably designed to be used as a show box. Unfortunately the lens is missing but some thirty engravings for viewing, mostly architectural views of European towns, accompany the apparatus. When closed, the camera obscura resembles a book bound in full calf leather, 21 × 14 × 6½ inches in size, and the spine is decorated with bands and flowers in gold tooling. The inscription 'OPTIQUES ET CHAMBRE OPSCURE PAR SEANEGATTIL' is poorly executed with a mixture of large and small capital letters and uneven spacing. 'Seanegattil' is assumed to be the maker's name.

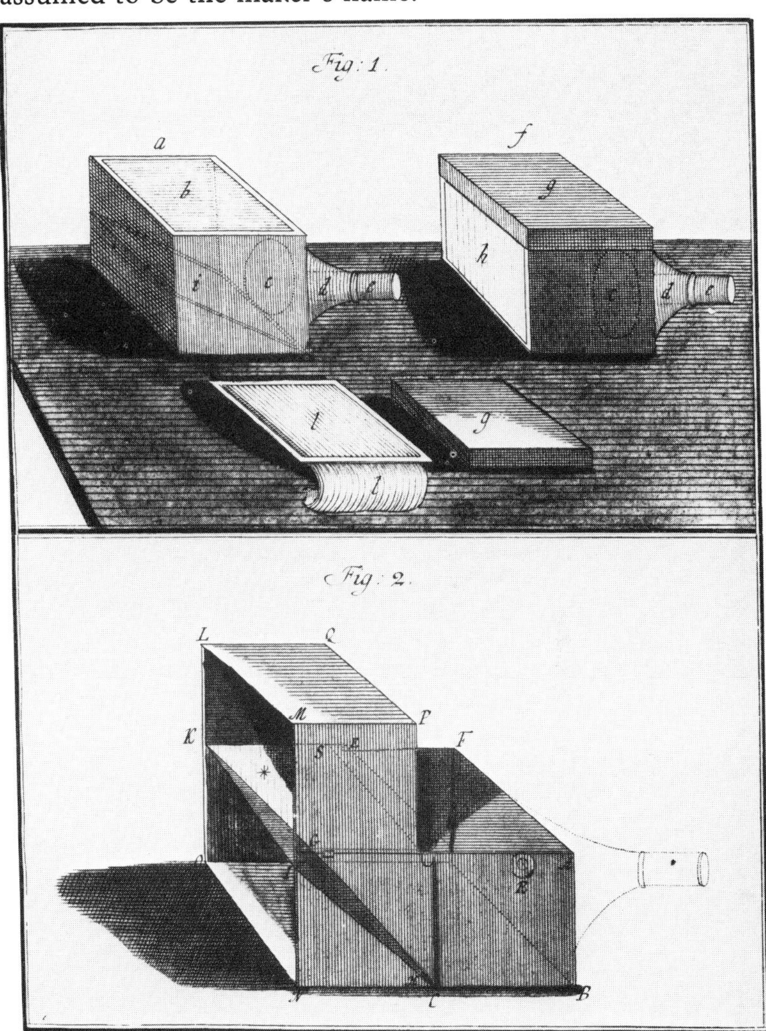

63. Detail drawing of Ledermuller's camera obscura attachments for a solar microscope, from his book on microscopy (1760).

64. A book-form camera obscura which could also be used as a show box. It was made in France in about 1750 and is now at the Victoria and Albert Museum, London. Crown copyright, Victoria and Albert Museum.

Harvard University has a very well preserved book-form camera obscura similar to that which belonged to Joshua Reynolds. A fire at Harvard in 1764 destroyed many of the scientific instruments and apparatus, although replacements were obtained from the London instrument makers John Dollond, Benjamin Martin and Edward Nairne. Benjamin Martin supplied the book-form camera obscura at a cost of £3 13s 6d, with an invoice dated 17 August 1765. He also supplied a camera obscura in the form of an artificial eye, consisting of a sphere of brass with a lens set on the surface and a ground-glass screen opposite.

Although few references to the use of the camera obscura in science have been found, there is no doubt that astronomers had continued to use it for observing sunspots.

Wind and cloud movements were usually recorded by observing cloud shadows on the ground, but J H Lambert (1728–77), the German physicist, used a camera obscura for the determination of the height of clouds when making meteorological records.[32] The National Museum of the History of Science at Leyden in Holland has two camera obscuras dating from about the middle of the eighteenth century, one of which is said to have been made in England.

An issue of the journal *Post Boy* of 12 March 1709[33] contained an advertisement for 'A curious original moving picture which came from Germany that was designed for the Elector of Bavaria,' and *The Tatler* of 29 December 1709 advertised 'A picture finely drawn, by an extraordinary master, which has many curious and wonderfully pleasing and surprising motions in it, all Natural.' Other similar advertisements appeared in journals during the early part of the eighteenth century, but it is uncertain whether they referred to camera obscuras. It is possible that the animated pictures could have been produced by a series of moving lantern slides or paintings in the form of relief panels with moving parts, such as the vanes of a windmill. Many ingenious devices and toys were made at the time, particularly for the wealthy courts of Europe.

EDWARD NAIRNE,

Optical, Philoſophical, *and* Mathematical Inſtrument-Maker,

At the GOLDEN SPECTACLES, REFLECTING TELESCOPE *and* HADLEY'S QUADRANT, *in* Cornhill, *oppoſite the* Royal Exchange, LONDON;

MAKES and Sells Spectacles, either of GLASS or BRAZIL PEBBLE, ſet in neat and commodious Frames, ſome of which neither preſs the Noſe nor Temples.

CONCAVES for Short-Sighted Perſons; READING, BURNING and MAGNIFYING-GLASSES.

Newtonian and *Gregorian* REFLECTING TELESCOPES; alſo EQUATORIAL TELESCOPES, or PORTABLE OBSERVATORIES.

REFRACTING TELESCOPES of all Sorts; particularly one of a new CONSTRUCTION (which may be uſed either at Land or Sea) that will ſtand all Kinds of Weather without warping, and is allowed by thoſe eſteemed the beſt JUDGES, who have made ſeveral late Trials of them at Sea, to exceed all others yet made in *England*; alſo a peculiar Sort to be uſed at Sea in the Night.

MICROSCOPES, either DOUBLE, SINGLE, AQUATICK, SOLAR, or OPAKE; likewiſe a new-invented POCKET MICROSCOPE, that may be uſed with the SOLAR, and anſwers the Purpoſes of all the other Sorts.

CAMERA OBSCURAS for delineating Landſkips and Proſpects (and which ſerve to view Perſpective Prints) made with truly Parallel Planes; SKY-OPTIC BALLS; PRISMS for demonſtrating the Theory of Light and Colours; CONCAVE, CONVEX, and CYLINDRICAL SPECULUMS; MAGICK LANTERNS; OPERA GLASSES; OPTICAL MACHINES for Perſpective Prints; CYLINDERS and CYLINDRICAL PICTURES.

AIR PUMPS, particularly a ſmall Sort that ſerves for Condenſing; AIR FOUNTAINS of various Kinds; GLASS PUMPS; WIND GUNS; PAPIN'S DIGESTORS; alſo a PORTABLE APPARATUS for ELECTRICAL EXPERIMENTS, which is allowed by the CURIOUS to exceed any of the Kind.

BAROMETERS, DIAGONAL, STANDARD, or PORTABLE.

THERMOMETERS, whoſe Scales are adjuſted to the Bores of their reſpective Tubes; HYGROMETERS; HYDROSTATICAL BALANCES, and HYDROMETERS.

HADLEY'S QUADRANTS, after the moſt exact Method, with Glaſſes, whoſe Planes are truly parellel; DAVIS'S QUADRANTS; GLOBES of all Sizes; AZIMUTH and other Sea Compaſſes; LOAD-STONES; NOCTURNALS; SUN-DIALS of all Sorts; Caſes of DRAWING INSTRUMENTS; SCALES; PARALLEL RULERS; PROPORTIONAL COMPASSES and DRAWING PENS.

THEODOLITES, SEMICIRCLES, CIRCUMFERENTERS, PLAIN TABLES, DRAWING BOARDS, MEASURING WHEELS, SPIRIT LEVELS, RULES, PENCILS, and all other Sorts of OPTICAL, PHILOSOPHICAL, and MATHEMATICAL INSTRUMENTS, of the neweſt and moſt approved Inventions, are made and ſold by the abovesaid *EDWARD NAIRNE.*

65. The trade card of Edward Nairne who made a camera obscura for James Bruce, British Consul in Algiers, in 1763.

A camera obscura made by Dominico Selva of Venice is illustrated in Coke's *100 Years of Photographic History,* and it is also mentioned by Schwarz, who noted that it 'can be exemplified by the camera used by Canaletto.'[9]

Fillip Nikitch Tiryutin, a Russian instrument maker, received commissions from several departments of the Academy of Sciences in Russia.[34] The drawing section of the Academy ordered 'two machines for taking perspectives' for Mikhail Ivanovitch Makhaev who, at the time, was an apprentice but later became an artist well known for his drawings of town buildings and street scenes of Moscow, St Petersburg, Kronshtadt and many other towns in Russia. The drawings were reproduced by engravers and were widely circulated. It is reported that no one knew how Makhaev made the perspective views until the documents relating to the instrument maker Tiryutin were discovered.

Illustrations in the paper about Tiryutin by Chenakal were taken from Middleton's *New Complete Dictionary of Arts and Sciences,*[8] and from *Leçons de Physique Expérimentale* by Abbe Nollet.[7] They are described as being typical of the period; presumably there are no engravings of Russian instruments.

Appendix

Gernsheim mentions a reference to a camera obscura in a carriage (in *Hannoverische Gelehrte Anzeigen* 1753, No. 44).

William Cheselden's book *Osteographia* was reviewed soon after publication in 1773 by John Belchier, in the *Philosophical Transactions of the Royal Society No.* 430, p194.

A book-form pyramid camera obscura made by George Adams is now in the possession of Mr Lionel Hughes of Coventry. Adams was a well known instrument maker in London during the latter half of the eighteenth century. When closed, the camera obscura measures 24 × 16 × 6 inches, and bears the legend 'Camera Obscura by Geo: Adams, Math. Instrument Maker' blocked in gold on the spine. When erected it is 33 inches high, the construction is of polished mahagony panels and a pine base, covered with black linen. The biconvex lens is 2⅞ inches in diameter with a focal length of about 24 inches, and it is fitted to a rack and pinion focusing movement of about 6 inches. A second lens is provided in a drawer above the mirror housing; this may have been a spare lens, since it has a focal length about the same as the one fitted, but more probably it was intended to supplement the fitted lens in order to provide a close focusing distance to allow for prints and paintings to be copied.

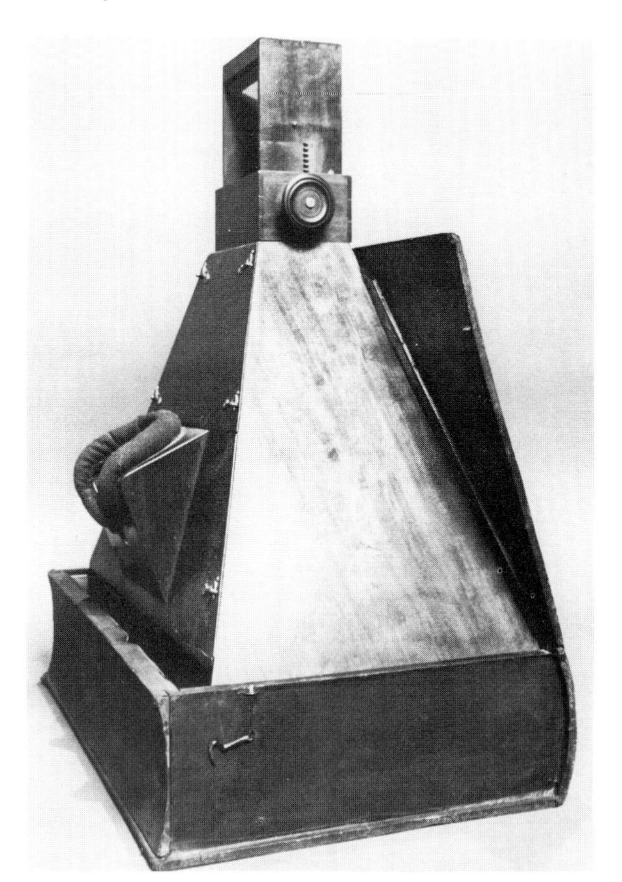

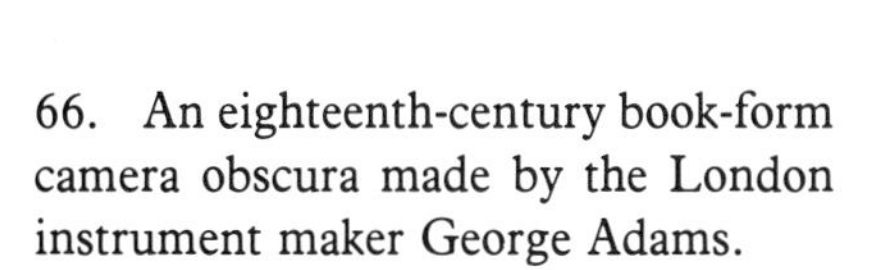

66. An eighteenth-century book-form camera obscura made by the London instrument maker George Adams.

References

1. Addison, Joseph 1712 *The Spectator* June
2. Quarrell W H and Mare M 1934 *London in 1710 from the Travels of Zacharias Conrad von Uffenbach* (London: Faber and Faber)
3. Science Museum 1971 *Scientific Trade Cards* (London: HMSO)
4. Gernsheim H 1969 *History of Photography* (London: Oxford University Press)
5. *British Journal of Photography* November 1916
6. Harris, Joseph 1775*Treatise of Optics* (London)
7. Nollet, Abbe 1755 *Leçons de physique Expérimentale* vol. 5 (Paris); 1752 *Académie Royale des Sciences* vol. 6
8. Middleton, Rev. E 1778 *New Complete Dictionary of Arts and Sciences*
9. Coke, Van Deren 1975 *One Hundred Years of Photographic History* (Albuquerque: University of New Mexico Press)
10. King H C 1955 *History of the Telescope*
11. Storer W M 1782 *Storer's Syllabus . . .* (London); 1845 Mechanics Magazine vol. 43
12. Walpole, Horace 1926 *Selected letters arranged by William Hoadley* (London: Dent)
    Whitwell, Elwin 1871 *The Works of Alexander Pope* (London: Murray)
13. Martin, Benjamin 1764 *Biographia Philosophica* (London)
14. Gill A T 1975 *Royal Photographic Society Journal* May
15. Martin, Benjamin 1740 *A New and Compendious System of Optics*
16. *British Journal of Photography* December 1916
17. Smith, Robert 1738 *Compleat System of Optics* (Cambridge)
18. Blunt R 1923 *Mrs Montague, Queen of the Blues: Letters and Friendships from 1762–1800* (London: Constable)
19. Brewster, David 1805 *Ferguson's Lectures* (Edinburgh)
20. Hutton, Charles 1803 *Recreations in Mathematics and Natural Philosophy* (London); 1796 *Mathematical Dictionary*
21. Guyot E G 1770 *Nouvelles Récréations Physiques and Mathématiques* (Paris)
22. Hooper Dr W 1787 *Rational Recreations* (London)
23. Harris, John 1704 *Lexicon Technicum* (London)
24. Gross, Harry I (nd) *Antique and Classic Cameras* (New York)
25. *Gentleman's Magazine* April 1753
26. Hinton, John 1752 *Universal Magazine* May
27. Priestley, Joseph 1772 *History and Present State of Discoveries Relating to Vision, Light and Colours* (London)
28. Nicolson M and Rousseau G S 1968 *This Long Disease, My Life: Alexander Pope and the Sciences* (Princetown)
29. Richardson, Samuel 1754 *Sir Charles Grandison* 1972 edn (London: Oxford University Press)

30. Sterne, Laurence 1781 *Tristram Shandy* (London: Dent)
31. Ledermuller M F 1760 *Microscopischer Gemuths und Augenergotzung* (Nuremburg)
32. Lambert J H 1773 *Nouveaux Mémoires de l'Académie Royale*
33. Victoria and Albert Museum 1972 *From Today Painting is Dead*
34. Chenkal V L 1953 *Russian Tool and Instrument Makers of the 18th Century* (Smithsonian Institution, Washington 1976)

The Nineteenth Century

> An instrument that has bestowed such incalculable benefits upon humanity.
>
> F Marion *Wonders of Optics* (1868)

There is little doubt that during the nineteenth century the camera obscura reached the height of its popularity as a useful technical device and as an entertaining diversion. The large room-type camera obscura became an attractive entertainment and many were built in gardens, parks and at holiday resorts. Technology was advancing rapidly and new uses were found for the portable camera obscura, but perhaps more important was the role that it played in the invention of photography. However, in spite of this simple means of making pictures, many amateur artists still preferred to sketch or paint and frequently made use of the camera obscura, which was consequently manufactured and sold throughout the century.

Photography

The invention of photography was made independently and almost simultaneously in 1839 by Louis J M Daguerre, a French artist, and W H Fox Talbot, an English landowner and scientist.[1-5] Some twenty years earlier, however, an etcher, Nicephore Niépce, had experimented with a form of photography which also involved a camera obscura. Daguerre used a small metal plate coated with light-sensitive silver halide; Talbot used silver halide on paper and Niépce used a sheet of metal coated with light-sensitive bitumen. The coated material was placed in the image plane of a camera obscura where it was exposed to the image and then processed (developed). At this stage each inventor employed different techniques, with the result that Daguerre produced a single finished picture, Niépce made a printing plate (the forerunner of photogravure printing), and Fox Talbot produced a negative from which many prints could be made. Talbot's method is that used in present-day photography.

Military Applications

During the 1860s a camera obscura was used for a defence experiment at Anvers in Belgium. It was set up to present a picture of the River Scheldt in which an explosive mine was sunk, and the position of the mine was indicated by a mark on the table of the camera obscura. Any hostile ship making its way along the Scheldt could therefore be seen, and as it approached the marked position the mine could be exploded by an electric current. The report of the experiment in the journal *Revue Maritime et Coloniale* of 1869 noted that the same method of defence had been used in Venice in 1859.[6] This report was said to have been an abstract of an article in an English journal *Mechanics Magazine,* although the present author has failed to discover the original.

In the book *Discoveries and Inventions* by Routledge, a chapter on torpedoes mentioned the Venice defence referred to in *Revue Maritime et Coloniale.*[7] The torpedo seems to have been simply a case of explosive material sunk in a waterway which was detonated when an enemy ship drew near. According to Routledge, self-activating torpedoes were unsatisfactory because 'the apparatus is liable to get out of order'; the best method appeared to be detonation by an electric current along wires from the torpedo to the shore. However, since the exact location of the torpedo was not known, there was the possible danger of blowing up one's own ships as well as those of the enemy. This problem was solved in Venice during the war with Austria in 1866. A camera obscura was erected to show the entrance channel to the harbour on the table. 'While the torpedoes were being placed in their positions an observer was stationed at the table, who marked with a pencil the exact spot at which each torpedo was sunk into the water. Further, those engaged in placing the torpedoes caused a small boat to be rowed round the spot where the torpedo had been placed, so as to describe a circle the radius of which corresponded to the limit of the effective action of the torpedo.' The course of the boat was then numbered and a corresponding number placed on a switch in the wires leading to that torpedo. When the image of an enemy ship entered the circle the appropriate switch was closed and the torpedo exploded. The chapter ended, however, 'The events of the war did not afford an opportunity of testing practically the efficiency of the preparations.'

Observatories

Kilmarnock

'In the interests of safety and slum clearance, however, the Tower had to be sacrificed.' So ended a brief note in the *Kilmarnock Standard Annual* of 1957.[8,9] All that remains of Morton's Observatory Tower, which he built in 1818 at a cost of £1000, is a newspaper reproduction of a water-colour painting; even the original painting is now lost. Thomas Morton (1783–1862), the son of a brickmaker, was apprenticed to a Mr Bryce Blair, turner and wheelwright, at about the age of ten. After a few years he started a business of his own and became very interested in scientific instruments. In this field Morton was self-taught through reading and by talking to stall-holders selling telescopes and barometers in the market place of Kilmarnock (a local history book noted that the salesmen were 'itinerant Italians').[10] Morton was particularly interested in telescopes and it was not long before he constructed one for himself. In his profession as a mechanic he made important contributions to the local carpet-weaving trade by designing and

67. Mr Morton's Observatory Tower at Kilmarnock, built in 1818 and demolished in 1957, in which two telescopes and a camera obscura were installed. This illustration, reproduced from a water-colour painting, was printed in the *Kilmarnock Standard* when the tower was pulled down. It is thought to be the only surviving picture of the Observatory.

making the 'barrel machine' and improving the Jacquard loom; in 1808 he was awarded £20 by the Board of Trade for his services to industry. Ten years later, at the age of 35, he built his Observatory Tower, some 80 feet high, in which he installed a Newtonian telescope of 9 feet 2 inches focal length and diameter of 9½ inches, a 7 inch Gregorian telescope and a camera obscura.

Morton's achievements as an instrument maker and optician were recognised by the people of Kilmarnock who, in 1826, gave a public dinner in his honour and presented him with a silver punch bowl. In 1835 the Royal Scottish Society of Arts presented him with an honorary membership. It was in the same year, or perhaps early in 1836, that he also installed a telescope and camera obscura at Dumfries. In the beginning of the twentieth century the Tower was used by a Mr James Kilmurray who was able to make remarkably accurate weather forecasts from astronomical observations.

Dumfries

The Dumfries and Maxwelltown Astronomical Society, founded in 1835, purchased an old provender mill which

68. The museum and camera obscura at Dumfries. The building was originally a flour or provender windmill built in about 1760, but the local astronomical society converted it to an observatory in 1836 and Mr Morton of Kilmarnock supplied a telescope and a camera obscura.

they converted to an observatory.[11,12] Mr Morton of Kilmarnock and a Mr Adie, an optician from Edinburgh, were invited to tender for the job of supplying a telescope and a camera obscura. Morton's tender was accepted and he was to supply 'The Camera of the size stated by you to be the most approved one viz., seven and a half inches, to be completely and neatly furnished with a circular table etc., and on the whole to be as good an Instrument if not better than the one now in the Edinburgh Observatory.' Morton charged £73 for the Gregorian telescope and £27 10s 0d for the camera obscura.

When the Observatory was opened to the public in July 1836, it was an immediate success and has continued operating to the present day. The camera obscura is now part of the Dumfries Museum which belongs to the local authority. About fifteen people can be accommodated round the table, which has a concave surface and may be raised or lowered for focusing. The lens is telescopic, giving a picture of about 1½ times magnification, and the mirror angle can be adjusted. The turret, which may be revolved, is glazed and has a sliding shutter for extra weather protection; on a clear

69. Interior of the Dumfries camera obscura.

day, landmarks some forty miles away on the horizon are recognisable. Thomas Carlyle, the philosopher and writer, signed the visitor's book several times during the first few weeks after the opening of the Observatory. In the following month, August 1836, a party of schoolboys from Burnside's Academy went to see the camera obscura and it is thought that this may be the first record of the visit to an institution of a school party.

Edinburgh

The camera obscura at Edinburgh has had a most tortuous history of which the first hundred years is told with humour and wit by J Willox in his small book *The Edinburgh Tourist and Itinerary* (1856).[12-17] It was proposed that Edinburgh should have an observatory fully equipped with instruments in as early as 1736, but at that time there was considerable social unrest from the Porteous riots and 'agitation and commotion of the people.' Consequently the project was shelved until 1741 when the Earl of Morton gave £100 towards the erection of an observatory. Professor Maclaurin and his colleagues at the university also made contributions, and the sum was brought to a total of £300. Five years later, in 1746, Professor Maclaurin died, and the two people holding the £300 unfortunately became bankrupt; consequently Dr Alexander Munro and Mr Short negotiated for the Observatory fund and recovered £400 by way of principal, interest and claims on the estates of the bankrupts. Willox referred to Short as the brother of the well known optician celebrated for his improvements to reflecting microscopes†. As a result of the efforts of Munro and Short, public interest in the Observatory was revived, an architect, a Mr Craig, was appointed and the foundation stone was laid in August 1776. 'But the measure of varying fortune to this seemingly fated institution was not yet filled. About this time Edinburgh was

†Probably the same Mr Short mentioned by Joseph Harris in the eighteenth century.

70. (*a*) The Outlook Tower, Edinburgh.

visited by Mr Adam, the architect.' Adam suggested that the Observatory should take the form and appearance of a fortress. 'But', wrote Willox, 'the inherent ugliness and inappropriate nature of the fabric were not its only or worst faults.' The construction very soon swallowed up all the funds, and the debt to the builder 'was ultimately liquidated by the successful issue of a horse race; and so the luckless Observatory was once more abandoned to its fate.' The town council eventually completed the building, which became known as the Old Gothic Tower, but they did not supply any instruments.

In 1812 an Astronomical Institution for Edinburgh was proposed and was established through the energy of Professor Playfair. The Institution had some influence with the town council who gave them a grant of the ground, the empty Observatory (the Old Gothic Tower) on Calton Hill and a 'Seal of Clause' which entitled the Institution to the privileges of a corporation. All these negotiations sound like a league of Town and Gown! The Institution then set out their objectives which were to be scientific and popular; they purchased instruments for the Old Gothic Tower, and in April 1818 building commenced. The Observatory was eventually opened to the public in 1856. In the meantime, however, Short had set up his own temporary observatory which contained a powerful telescope and a camera obscura which was 'kept by a female descendant of the celebrated Short, already spoken of.'

These details are confirmed in a pamphlet dated 1846 entitled 'Some remarks on the present state and future prospects of the Observatory of the Astronomical Institution of Edinburgh.' The old Observatory building on Calton Hill was referred to as the 'Old Gothic Tower' in the pamphlet, which also said that the Institution was granted the use of the Tower and 'gave orders to Troughton to prepare some first-rate Astronomical Instruments for the Scientific Observatory about to be built, and bought some minor Instruments at once for the observatory in the Gothic Tower . . . Afterwards a Camera-Obscura was erected in the upper part of the building. [This would be about 1818 according to Willox] The Scientific Observatory . . . was almost completed when George IV visited Edinburgh. The Members presented a loyal and congratulatory Address to His Majesty, who was pleased, in return, to grant them permission to call their Institution "The Royal Observatory of King George IV".'†

When the building was completed in 1823 or 1824, after the Royal visit, the Institution had no money left to obtain apparatus and this state of affairs went on for some five or six years. The pamphlet continued:

> The funds of the Institution having been expended on the buildings, the minor instruments, and the servants, and exhibitors of the Camera, the Government, in 1830, granted £2000 for the purchase of instruments for the Scientific Observatory . . . At length, in 1834, under the Presidentship of General Sir T. Makdougall Brisbane, Bart., the Institution made over to the Government the unlimited use of the Scientific Observatory, on the condition that the Government would always keep a Salaried Astronomer . . .

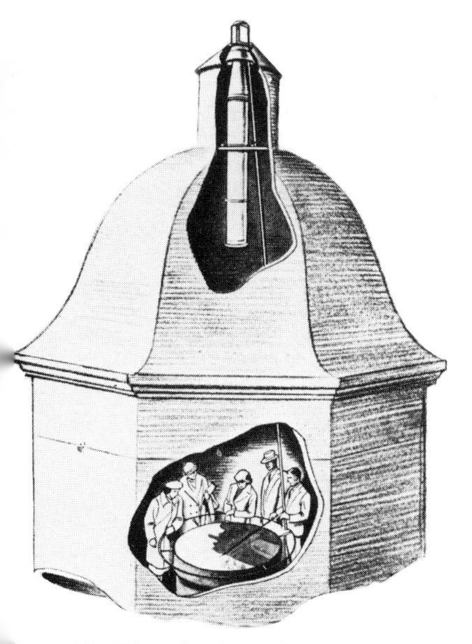

70. (*b*) The Outlook Tower camera obscura at Edinburgh. An illustration from a catalogue of Messrs Barr and Stroud who fitted new lenses and a mirror in 1955.

It has not been possible to discover exactly how long Edinburgh had these two camera obscuras, but in 1856 Willox referred to a camera obscura at the 'old building' (the Gothic Tower) and to one at Short's Popular Observatory, both of which were on Calton Hill. After that, dates and movements are somewhat confused, although it seems that one camera obscura (probably Short's) ceased to exist and the other was moved from the Gothic Tower to the Outlook Tower on Castle Hill. The installation of a camera obscura in the Outlook Tower took place before 1892, the year it was acquired by Sir Patrick Geddes the town planner. In 1947 the camera obscura was overhauled and fitted with new apparatus by Messrs Barr and Stroud of Glasgow, and in 1955 a warrant was granted to construct a glass house to accommodate the original equipment. During the same year, the 1947 model in the Outlook Tower was replaced by a modern lens and mirror which, it was said, gave 110%

†Against this last paragraph has been pencilled '1822'.

more light. A Barr and Stroud catalogue of the late 1960s contained a draughtsman's sketch titled 'C.G. 3 Camera Obscura, fitted in the Outlook Tower, Edinburgh'. In 1970 the Tower was bought by the University from the Church of Scotland, the camera obscura was retained, and is now a tourist attraction.

Glasgow

The city of Glasgow also acquired a camera obscura in the nineteenth century. The Society for Promoting Astronomical Science had begun the building of an observatory in 1810, but it is uncertain whether it was ever completed.[18] The Observatory was to serve three functions: members of the Society were to have the facilities of a clubroom and reading room; some accommodation was to be set aside for serious scientific study and observation, and the third portion was to house a camera obscura and telescopes for use by the public and for popular observations. It appears the Society was not able to finance the project or the building, if it was ever completed. A request for finance was made to the university but this was refused, and so the Observatory foundered. Although Brewster's *Dictionary* of 1830 refers to a camera obscura at Glasgow it may have ceased to function by that year.

Stranraer

Sir John Ross, the Arctic explorer, built a house at Stranraer which he called North West Castle.[19] At some time between 1820 and 1824 he installed a camera obscura which remained in working order throughout the century until the house was taken over by the Royal Air Force in 1940. The apparatus was then put into store with a number of other instruments relating to Sir John's voyages. No further information is available and nothing is known at Stranraer of the present whereabouts of the apparatus.

Hull

The Zoological Gardens at Hull were set out in 1840 under the direction of a Mr Lockwood, an architect who also designed the buildings which included a camera obscura, although nothing now remains.[20] By 1860 or so all the buildings were demolished to make room for houses to accommodate the fast-growing population of the Spring Bank area.

Liverpool

During the 1830s Mr E Henderson was the proprietor of an observatory at Mount Gardens, Liverpool.[21] It contained many instruments, demonstrations, maps, diagrams, a camera lucida and a camera obscura 'for terrestrial and celestial views'. Henderson styled himself 'Professor of Astronomy', and wrote several small books on the subject of celestial observation. In a pamphlet on mathematics by Henderson an advertisement page announced his courses of lectures on astronomy which he delivered at local academies, seminaries and at his observatory.

Llandudno

On a hill overlooking the Happy Valley at Llandudno, Mr Lot Williams erected a camera obscura in 1860 which, although only a timber building, remained sound for over a hundred years.[22,23] The camera obscura soon became a popular amusement and the hill on which it stood became known as Camera Hill. It was a family possession but in 1964 the owner, Mrs R G Jones of Conway, surrendered the tenancy to the local council, and the camera obscura was then closed. Unfortunately, in October 1966 the building was broken into by vandals who started a fire, and the entire structure was destroyed. Very little information is now available about the Llandudno camera obscura, although it is known that the room could accommodate about a dozen people around a fixed table approximately four feet in diameter. Sharp focus of the image was achieved by moving

71. The prospect tower at Jersey Marine. The top floor was roofed with a cupola which contained a camera obscura. The tower was built, with an adjacent hotel, in about 1850. This photograph was taken in the autumn of 1979.

the lens with a control rope, while another rope adjusted the mirror angle and a double rope rotated the turret containing the lens and mirror. It is unfortunate that this camera obscura, as well as others which have suffered a similar fate, have not been replaced. Reconstruction would not detract from the historical aspect of a camera obscura and it would be, as it was in the past, a fascinating entertainment, particu-larly in holiday resorts.

Jersey Marine, South Wales

In the middle of the nineteenth century an attempt was made to develop a seaside resort at Jersey Marine, a small village near Neath in South Wales, although it was not a success because the prevailing winds carried unpleasant fumes and soot from the smelting works and heavy industry at Swansea.[24,25] However, John Taylor, a local architect, built a hotel and a five- or six-storey octagonal prospect tower. At the top of the tower he constructed a cupola and installed a camera obscura. The hotel was demolished in 1968 and although the tower remains it is somewhat dilapi-dated; no one knows when the camera obscura disappeared. Barbara Jones wrote about Jersey Marine in *Follies and Grottoes* (1974) and said the tower was being used as a stable, although more recent information has reported it as a bottle store for a local brewery.

Bristol

One of the few remaining permanent camera obscuras can be found at Bristol on Clifton Down overlooking the Sus-pension Bridge;[12] it has already been mentioned as William West's creation from an old snuff mill which had remained derelict after a fire in 1777.[26,27] The property was owned by the Society of Merchant Venturers who leased it to West for five shillings a year. Latimer in his *Annals of Bristol* (1887)[2] said that Mr West had fitted it with a telescope and a camera obscura in 1829 and called it an observatory, a title it has

72. The Clifton Observatory and camera obscura at Bristol, installed by William West in 1829. Originally the building was a snuff mill but a fierce gale in 1777 turned the sails so fast that the bearings caught fire and the mill machinery was completely destroyed.

OBSERVATORY, CLIFTON.

Mr. WEST begs most respectfully to inform the Subscribers, Visitors, and the Public generally, that the OBSERVATORY is now open, and that it contains, for their use, by day or night, large and powerful

REFLECTING AND ACHROMATIC TELESCOPES;

ONE OF 20 FEET,

And several of smaller dimensions for astronomical purposes, and for viewing the extensive and beautiful scenery by which the Observatory is surrounded.

Upon the summit of the Tower is placed an excellent and unusually large

CAMERA OBSCURA,

(The diameter of the picture being 5 feet), embracing the whole of the extensive view from the base of the Tower to the horizon. The Camera Obscura, to those unacquainted with it, has a magic effect. The entire of the varied scenery within the range of the summit of the Observatory, including figures, and animals in motion, carriages, the waving of foliage, ships and packets arriving and departing, being brought in succession into the picture with the distinctness and vivid colouring of nature, and affording a high gratification to the observer from the continued changes and varied effects of light and shade upon the landscape.

A variety of amusing and pleasing OPTICAL INSTRUMENTS, EPPERIMENTS, &c.

GALVANIC APPARATUS.

Every description of PHOTOGENIC DRAWING shown and explainad.

Beautiful Specimens of the Daguerréotype.

73. William West's handbill announcing the completion of an extension to the Clifton Observatory in the summer of 1837. The new building cost over £1300 which was defrayed by donations, subscriptions and a grant from the Society of Merchant Venturers.

retained to the present time. William West died in 1861, but members of the family continued to live on the premises until 1943.

Plymouth

A camera obscura was built on Plymouth Hoe in 1827 by a Mr Sampson, although it was removed about 1891 to make way for the building now known as the Belvedere. Andrew Cluer, in his book about Plymouth reproduced two photographs (dated about 1865), of the camera obscura, which appears to have been an octagonal building with a shallow, sloping roof supporting a revolving turret containing the lens and mirror.[29] The room was probably ten or twelve feet in diameter and the height to the lens about twelve feet. The camera obscura is shown on a street plan of Plymouth by W H Maddock, dated 1877.

Ramsgate

A guide to Ramsgate for 1871 mentioned a camera obscura on the west pier of the harbour.[30] The present librarian at Ramsgate, Mr Busson, remembers it as a stone building with a red tiled roof accommodating about twenty people around the table. The camera obscura continued to operate for a few years after the last war, but it was then demolished to provide room for cars awaiting export and the optical apparatus was put into store.

Brighton

There may have been two camera obscuras in Brighton during the early nineteenth century, but the story is disjointed and somewhat confused.[31-37] The *Brighton Herald* of 5 September 1807 reported 'that most complete optical machine, the Camera Obscura situate at the bottom of the Steine, near the sea, was visited by the Rt. Hon. Earl Bathurst, Mr. Wilberforce and many others of the nobility . . . who declared themselves much satisfied with the pleasing exhibitions displayed therein.' According to Baxter's *Stranger in Brighton* directory of 1824 the camera obscura 'Is situated at Hayne's Toy Shop, on the Steyne-beach, and adjoining Russel House'; an entry in the directory section said 'Haines, Edward (royal camera obscura) Old Steine.' A guide book of 1838 referred to a camera obscura at the end of the esplanade, but there was no mention of Mr Haines. However, it seems quite certain that a camera obscura existed in the late 1830s and had been established since 1807.

In August 1824 the proprietors of Brighton Suspension Pier held a meeting at which Captain S Brown proposed to erect a camera obscura and cosmorama at the outer end of the pier. The buildings were to occupy not more than 10 feet by 12 feet each and for these he was prepared to pay a rent of £40 per year. The proposal was apparently temporarily shelved, but two years later a minute book entry for August 1826 read: 'Captain Brown having applied for leave to plan the Camera obscura between the chains over the Library, it is resolved that he be permitted forthwith to do the same.' Eighteen years later, in September 1844, the proprietors 'resolved that the camera obscura as erected over the Pier Lodge Apartments and let to John Gurr at twelve pounds per annum be purchased by the Company of Sir Samuel Brown for the sum of Two hundred pounds.'

By 1862, however, it seems that not all was well at Brighton; in *The Beauties of Brighton* Francis Coghlan wrote: 'We have nothing more to say on the subject of the Chain Pier, but that it might be made more attractive as well as remunerative, if the directors (who are they?) were to cater a little more for the amusement of the public; a camera obscura at the descent of the Marine parade, and also a Bazaar where you pay through the nose for every toy.' A quotation from a small book on the chain pier written by John George Bishop in 1897, may provide a suitable end to this saga. 'The Camera Obscura, which for several years had done duty on the Steine beach, immediately east of Russell House, was also placed at the Pier head' (later on this was removed to its present position, immediately over the Saloon, with an entrance to it from the Marine Parade). So perhaps there was only one camera obscura at Brighton after all, and it was moved around with each change of tenant.

74. An illustration from a guide book to Brighton by J Whittemore (1825).

Whittemore's guide to Brighton of 1825 is illustrated with an engraving showing a camera obscura on the pier, and the Brighton library has an undated photograph of a camera obscura which appears to be 'over the Saloon'. Lindley has nothing to add to the Brighton story.

Text Books

The language and style of Dr Neil Arnott's book, *Elements of Physics, or Natural Philosophy Written for General Use*, are far more interesting and amusing than the information on the camera obscura.[38] In fact, 'Language is wanting to express the momentous consequences to man of the power of a lens . . . With what rapture does the schoolboy first view this enchanting picture drawn by nature's own pencil, with colours taken directly from the sun's bright ray!' When referring to images made by small holes, Arnott related, 'Barry, the eminent painter, while lying on a sick bed, in fever, mistook such a scene appearing on the ceiling of his room for a supernatural vision.' Arnott was referring to the same James Barry who, for the cost of materials only, made

75. The tent camera obsc[illegible] popular among amateur artist[illegible] manufactured throughout the [illegible]teenth century. The turret could be revolved to present a different scene, and the mirror angle adjusted for near or distant views. (From *Natural Philosophy* by E Atkinson 1900.)

the paintings which surround the lecture theatre at the Royal Society of Arts.

In his book *Natural Philosophy*, Dr E Atkinson used somewhat more prosaic language to describe images produced in a darkened room which, he noted, were first observed by Porta.[39] Several editions of the book were published, but each one contained a different illustration. One of the pictures showed a tent camera obscura in use, and another depicted a darkened room in a house with two children looking at the image of a garden. However, the latter scene was misrepresented by the artist: the room is too high for the garden scene to have been projected, and the lens is too far inside the room; as illustrated, the set-up could not possibly have worked.

Professor Dionysius Lardner, of University College, London, wrote in his *Museum of Science and Art* a very

76. An illustration from *Natural Philosophy* by E Atkinson (1900). The hole in the wall is far too small for the wide-angle view shown on the table. No doubt this was due to the engraver's not understanding the subject of the illustration, a common fault in popular textbooks of the last century.

77. This camera obscura, made by Ross of London in about 1880, was covered with coach hide so that it looked like a travelling case. It was probably a special order because at that time the firm Ross was busy making photographic cameras. The lens of this camera has a set of Waterhouse stops which were typically used with lenses for photography.

simple explanation and description of several camera obscuras.[40] The chapter is well illustrated and he says 'This is an instrument of extensive utility in the art of design; by it the process of drawing is reduced to that of mere tracing.'

Dr David Brewster, in his *Treatise on Optics* (1831), observed that the camera obscura was 'an amusing and useful optical instrument' and that the 'perfect resemblance of the picture to nature astonishes and delights every person, however often they may have seen it.'[41] He also added a very useful suggestion, 'In the portable camera obscura I find that a film of skimmed milk, dried upon a plate of glass is superior to ground glass for the reception of the images.'

78. A camera obscura of about 1850. The ground-glass screen and mirror were mounted in a draw box which when closed made the instrument more compact for carrying. Focusing was achieved by means of a rack and pinion which moved the lens in a tube mount.

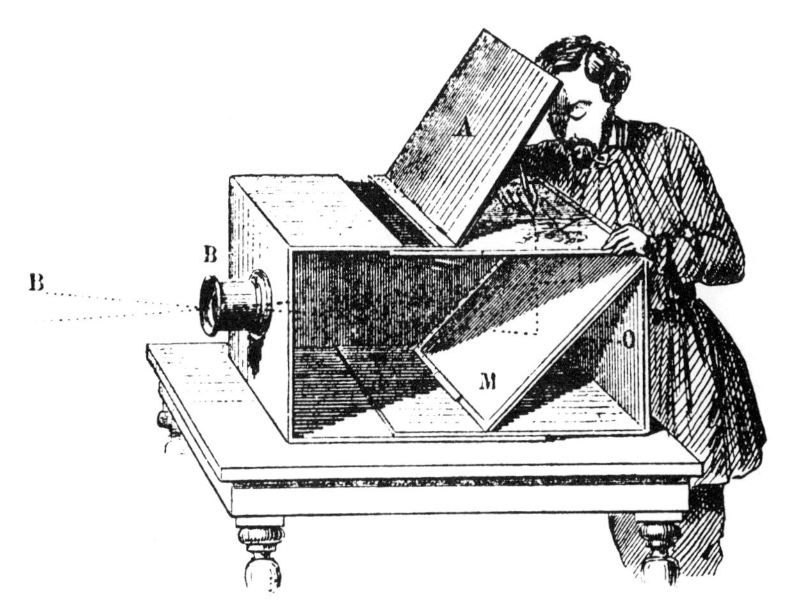

79. A box camera obscura from Lardner's book *Museum of Science and Art* (1855). This subject was frequently illustrated in books and encyclopaedias of the nineteenth century. There is a difference in scale between the camera obscura and the figure similar to that in the engraving of Kircher's camera obscura in the seventeenth century (p26).

In *The Forces of Nature* by Guillemin[42] is an illustration of images of the sun made by small gaps in the foliage of trees†. The picture shows a large number of white spots on the path of a leafy glade (see Frontispiece). These images may be seen in nature and are distinguishable from patches of sunlight by their almost uniform size and shape. The sizes will be different as the distances of the small openings vary, and the shape will be round or elliptical according to the angle of the ground to the sun's rays. The images will not be as bright as a patch of sunlight from a large gap in the foliage. To produce an image of the sun the 'openings in foliage' would have to be about an inch or less in diameter. Guillemin's other illustrations of camera obscuras (both room and tent types and of a megascope) are also drawn in a naturalistic style and the text is very simple.

In 1821 J B Biot described and illustrated a camera obscura and a megascope with which he said it was possible

†Since writing, the author has been shown a photograph of sun images in *The Universe of Light* by W H Bragg (1933).

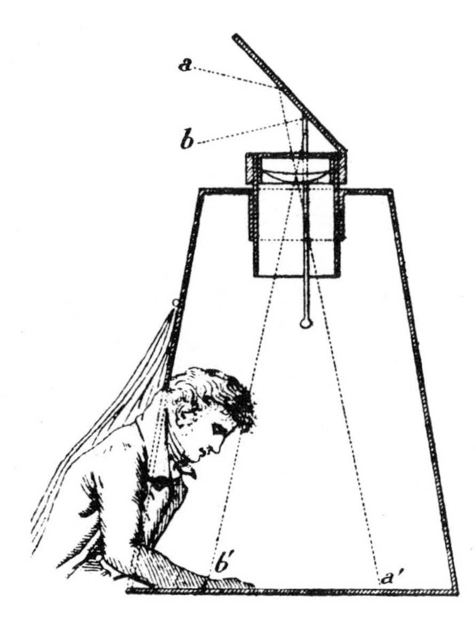

80. A pyramid camera obscura with a meniscus lens in a focusing mount and an adjustment for the mirror. The image is the right way up for the viewer. These instruments were usually made of mahogany and stood about two feet high. As in the previous illustration, the figure has been drawn to a reduced scale. (From a French dictionary of arts and crafts, 1823–35.)

to make images of from two to twenty times life-size enlargement.[43] He also referred to Wollaston's suggestion that a meniscus lens could improve the camera obscura image.[44] The French optician, Charles Chevalier, also made advances in lens design for the camera obscura.[45] In 1819 he replaced the lens and mirror combination (as used in a tent or a permanent camera obscura) with a prism which had lens surfaces. In 1823 he made a meniscus prism which increased the area of sharp focus as did Wollaston's meniscus lens. The single unit comprising lens and prism was much easier to mount than the separate lens and mirror, but as the angle of a prism cannot be adjusted it was not possible to bring near objects into view.

According to Marion in his book *Wonders of Optics* (1868) John Baptista Porta observed the image-making properties of small holes: 'This was the first attempt at the formation of a camera obscura, an instrument that has bestowed such incalculable benefits on humanity.'[46] The text is adequate and the illustrations are pleasing, particularly one of a river and castle scene in which four children are looking at the table of a camera obscura in the foreground. The camera obscura has a domed roof and looks very substantial, although the text says that it 'may be easily and cheaply constructed.'

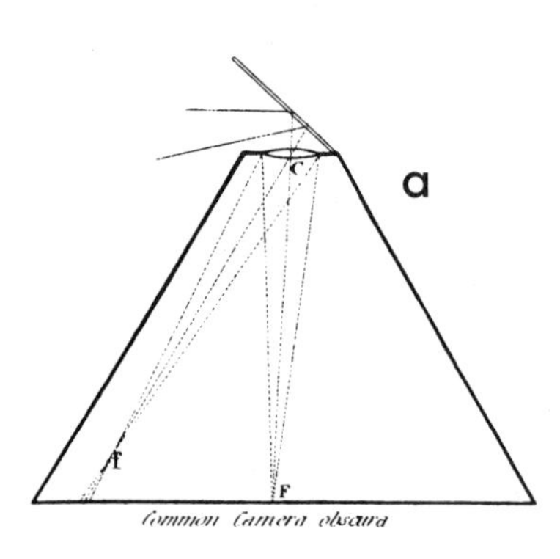

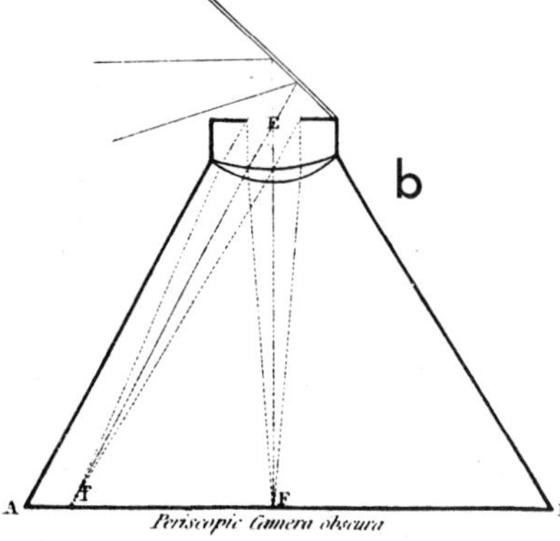

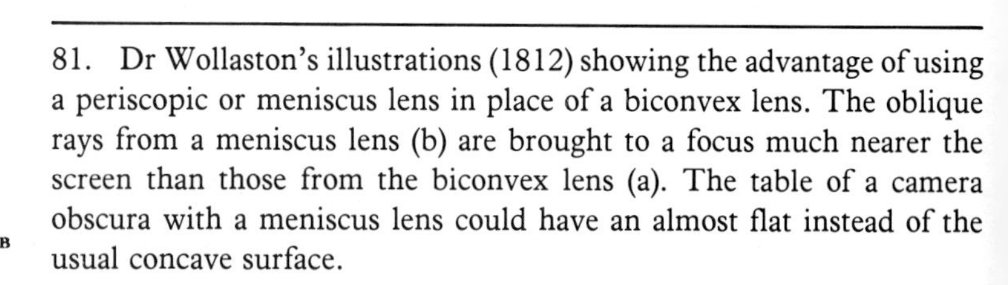
81. Dr Wollaston's illustrations (1812) showing the advantage of using a periscopic or meniscus lens in place of a biconvex lens. The oblique rays from a meniscus lens (b) are brought to a focus much nearer the screen than those from the biconvex lens (a). The table of a camera obscura with a meniscus lens could have an almost flat instead of the usual concave surface.

82. A French reflex camera obscura of about 1800, now at the Museum of Applied Arts and Sciences, Sydney, Australia. Its dimensions are 13×10×10½ inches, and the drawing area is about 6½×12 inches. A lens is mounted on a box which slides into the body of the instrument, and is the means of focusing. The construction is of cedar, and the acanthus leaf design of the lens box and lid are carved in mahogany. The top surface of the lid is inlaid with pearwood, and the underside is covered with a sheet of glass decorated with a gold design, and the ring handles are of ivory.

J Imison's *Elements of Science and Art* (1808) gave the usual description of image formation and about the use of a camera obscura for drawing.[47] He suggested making solar observations by replacing the lenses of a scioptric ball with a telescope so that it could be moved to follow the sun. The image, he said, could be thrown onto a white screen; 'In this manner the sun's face is viewed without offence to the eyes.'

In his *Histoire des Mathématiques en Italie* (1838) Gugliemus Libri-Carucci della Somaja mentions the camera obscura and appears to have been very much concerned with the priority of the 'inventors'. Although a Florentine nobleman he was notorious for stealing valuable literary works, which earned him the nickname 'Libri the book thief'.[48]

83. This picture of Victorian childhood comes from *Wonders of Optics* by F Marion (1868), who said that the camera obscura was 'an unending source of amusement and may be easily and cheaply constructed.'

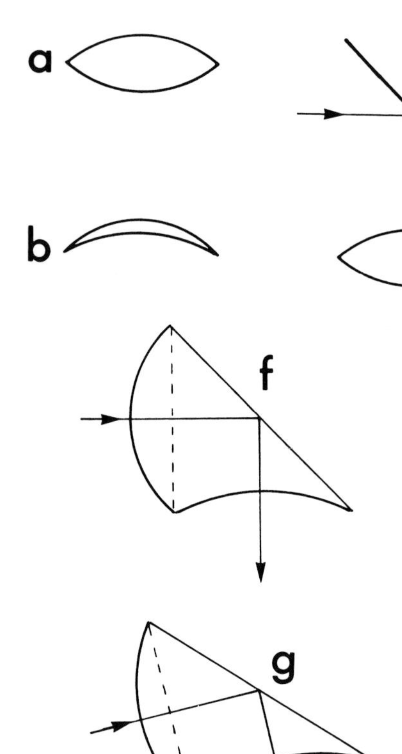

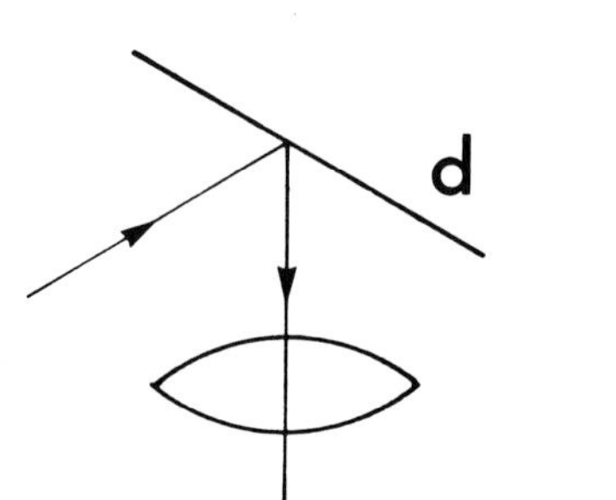

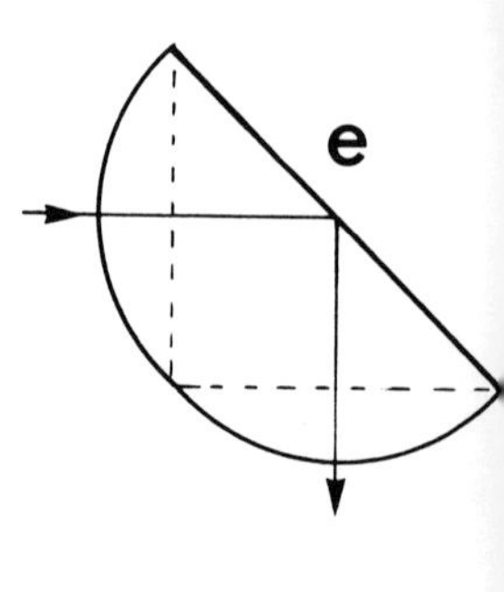

84. Lenses, mirrors and prisms.
(a) A biconvex lens.
(b) A meniscus lens, convex and concave surfaces.
(c) A mirror at 45° to the horizontal reflects distant objects projected by the lens.
(d) A mirror tilted at an acute angle to the horizontal reflects nearby objects. The angle of a ray reflected from a mirror is always equal to the angle of the incident ray.
(e) A right-angled prism with convex lens faces.
(f) A right-angled prism with convex and concave lens faces.
(g) A prism tilted to view a nearby object. The reflected ray is not vertical, as from a mirror, and consequently the image would not be formed on the table of a camera obscura. In a right-angled prism the reflected ray and the incident ray are always at 90°to each other.

Encyclopaedias

Several dictionaries have commented on the association of the camera obscura with the invention of photography. An *Encyclopaedic Dictionary* of 1882 noted: 'Prior to the discovery of photography the camera obscura was only an elegant scientific toy; but the important part it was made to play in that beautiful art at once raised it to the rank of a philosophical instrument.' The *Imperial Dictionary of the English Language* (ed. Ogilvie 1898) described a portable camera obscura 'which is the variety used by photographers', while *Knight's Practical Dictionary of Mechanics* (1877) contained a brief but informative history of the camera obscura and stated that it was used for photography.

However, 'before that art attained any celebrity' it was sold in stores and used for sketching from life and nature. The *Century Dictionary* (ed. Whitney 1889), published in New York, contained a brief entry in which it said there were various forms of the camera obscura 'but the design of them all is to throw the images of external objects, as persons, houses, trees, landscapes, etc., upon a plane or curved surface, for the purpose of drawing, the making of photographic pictures or mere amusement.'

The *Penny Cyclopaedia* (1836) gave an intriguing explanation for the camera lucida and camera obscura: 'In the first contrivance there is no *chamber*, but as it was the last invented, and as its predecessor had been called *camera obscura*, it was called camera lucida.' The article referred to Porta's description of a camera obscura and to his secret of placing a convex glass in the hole, but 'he does not appear to have found out that the screen should be curved, in order that the points of convergence of all the pencils [of light] should fall upon it.'

Metropolitania (1845) was somewhat phlegmatic about the invention of the camera obscura; it 'has been ascribed to Baptista Porta . . . But Dr. Freind, in his "Historie of Physic", observes that Friar Bacon . . . describes this instrument and various glasses . . . In fact, it seems highly probable that a similar effect . . . could not remain long unnoticed by optical writers or observers , as it is produced in a certain degree in every dark room in which a hole is found sufficient to admit a small pencil of rays, without the assistance of any lenses whatsoever.' This introduction was followed by an account of image formation and descriptions of three types of camera obscura: the box type was 'the most portable and is the form most frequently used by artists, on account of its convenient dimensions.' The article also made an interesting note on drawing: 'The image thus formed may be readily traced on the rough surface of the glass by a blacklead pencil, or by what is preferable, red French chalk, the white paper being gently placed on the glass the lines will be correctly taken off.' Also mentioned is the 'ingenious camera obscura, invented by Mr. Thompson, of Duddingston, which is so contrived that it may be shut up and carried like an umbrella.'

Of all the cyclopaedias, the one compiled by Abraham Rees in 1819 contained the most comprehensive article on the camera obscura, although the history was very brief and was concerned only with Bacon and Porta.[49] When describing the camera obscura Rees said

> . . . the use is manifold; it assists very much in explaining the nature and rationale of vision, and hence by some it has been compared to the artificial eye. It exhibits the most striking and entertaining representations of objects of all descriptions, whether near or distant, in their true perspective, the colouring just and natural, their light and shadows correct, and all their motions and relative positions according to the original. By means of the instrument, a person however unacquainted with drawing, may delineate objects with great facility and correctness; and to the skillful artist it will be found indispensibly useful in comparing his sketches with the perfect representations given in the camera, and by observing his defective imitations, he may correct, as much as possible, his designs. To the delineators of that beautiful representation called the Panorama, this instrument has proved of essential use.

Rees continued with the theory and construction of the camera obscura, and a method for correcting the image, as 'the inverted position of the images may be an objection'. He then discussed the scioptric ball and its uses for observing sunspots, and 'At the time of a solar eclipse the whole progress of the moon, from the time of the first contact of the limbs to the last, may, in this way, be observed very distinctly.' The following sections on permanent and portable camera obscuras are acknowledged as contributions from Mr William Jones, optician, Holborn. At the conclusion of an account of a reflex camera obscura Rees wrote 'Some artists who take profiles take out the rough glass from its cell, invert the camera, and by a stand support it about 10 or 12 inches above the white paper on the table. The image will then be invertedly formed on the paper, and they trace it with a pencil in a correct manner, and with less trouble than by any other method. Messrs Jones, of Holborn, make an improved camera of this kind.'

Storer's delineator was also mentioned, particularly his use of a lens immediately below the ground glass, 'but without the least pretensions, it being previously well known by the most eminent opticians, and it was, in the year 1758 noticed by Mr. Hooper, in his *Rational Recreations* Vol. XII p 29.' A similar comment was made by Good in *Pantalogia*, although the statement is not true. Hooper, when discussing Guyot's camera obscura, suggested that if the 'upper surface of the glass (which he referred to as clear) were convex it would be better still'—better meaning brighter. While making a picture this arrangement would be unsuitable for drawing because a sheet of paper would not conform exactly to the convex surface. Storer placed his plano-convex lens underneath the ground-glass screen (the plane side in contact with the ground glass) or under a sheet of clear glass, on which a piece of paper was placed for drawing. An image bright to the edges and corners of the screen or paper would thus be formed. It was mentioned in the previous chapter that Storer's principle is now adopted in modern single-lens reflex photographic cameras, except that the heavy and bulky plano-convex lens is replaced by a Fresnel lens.

The illustrations in Rees' *Cyclopaedia* were as clear and as informative as the text, showing a darkened room with an image made by a small hole, a scioptric ball with a solar mirror, and a structural camera obscura in which the dome roof and the turret could be revolved separately. A turret was also shown attached to the ridge of a pitched roof. Rees included illustrations of a reflex camera obscura, as well as a pyramid-shaped book-form camera obscura with an accessory for converting it to a show box for viewing prints.

Journals

The *Magazine of Science* commenced publication on Saturday, 6 April 1839 with an illustration of a camera obscura on the cover, showing a permanent structure about twelve feet square similar to a summer house.[50] The roof is about fifteen feet high with a turret containing a lens and mirror, and the table in the room appears to be a little over-sized, probably in order that the engraver could conveniently illustrate the image of a country scene. The text noted that the camera obscura 'is of such simple construction as to be easily understood, and represents the objects subjected to it with so unerring a fidelity, and in all their vivid colours that it has always been a favourite amusement to view its varied and animating pictures.' The article concluded, '. . . or it may be made still more easily in that form usually called a Portable Camera, a description of which will appear in our next Number.' This next description, however, was very short, without a diagram or dimensions, and ended: 'The Box Camera has lately come into more than ordinary notice, in consequence of being the instrument employed, both by Mr. Talbot and M. Daguerre, in their newly-discovered and important process of Photogenic Drawing.' A week or so later another article on photogenic drawing appeared, in

THE CAMERA OBSCURA.

85. This engraving appeared on the cover of the first issue of the *Magazine of Science*, 6 April 1839. The accompanying text explained the principles of image formation and the advantages of using a meniscus lens in place of a biconvex lens.

which the construction details and a diagram were given for the box camera obscura.

In 1842 the *Magazine of Science* published a letter describing an improved camera obscura for photogenic drawing. The single word 'camera'—meaning a photographic camera—had not yet come into general use. In the same year another letter gave precise instructions for making a pocket camera obscura with many hinges at the sides and ends which enabled it to be folded flat; when unfolded for use it became a box about 2×2×3 inches. Although this camera obscura was intended for the photographic process the letter submitted to the magazine has such charm that it is worth quoting:

> Sir—I hereby transmit to you the sketch, with description, of a newly constructed camera obscura, more portable than any I have yet heard of, and if worthy of a place in your columns, insertion in them would be conferring an obligation on—AEGIDIUS. Ipswich, April 1842.

In 1845 the same magazine reported on a patent granted to Mr W E Newton, a civil engineer of Chancery Lane, London. The patent described a copying machine which Newton claimed to be a new application of the camera obscura. The machine was 'for facilitating and copying of

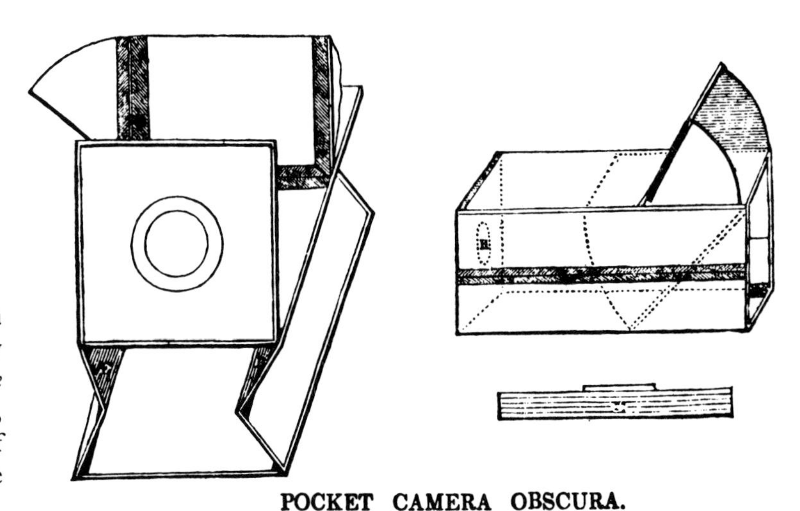

86. A miniature folding camera obscura intended for photography illustrated in the *Magazine of Science* (1842). It was 2×2×3 inches in size, and 'the hinges may all be made of strong holland, of course, glued to the wood.'

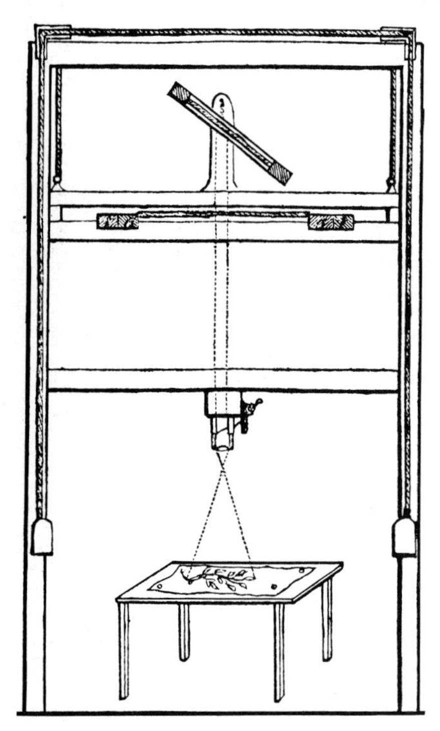

CAMERA OBSCURA
COPYING MACHINE.

87. This forerunner of the modern drawing office projector was described in the *Magazine of Science* (1845). It was used to enlarge or reduce drawings, and the original normally had to be made translucent but the patent specification describes a method for copying opaque originals.

designs, drawing, and etchings of all kinds, either of the original size, or upon enlarged or reduced scale.' The apparatus stood about nine feet high, consisting of four uprights between which a horizontal framed glass sheet could be moved up and down by cords and counterbalance weights. The drawing to be copied was oiled to make it translucent, and was placed on the glass sheet. Light (from an unstated source) was directed down through the drawing by a mirror at an angle of 45°. Below this mirror was a board with a lens in a focusing mount which was also counterbalanced for movement up and down. The lens projected an image of the drawing onto a small table placed between the uprights. The modern drawing office projector, such as the Halco–Sunbury 'Copy-scanner', is similar to Newton's invention; the only essential differences are that the original drawing is now illuminated from the front and the image is projected upwards.

Louis Spohr, a violinist and composer, kept a journal when travelling in Italy with his wife and children during the 1880s, in which he recorded, 'In a second vehicle which accompanied us travelled an Englishman who was possessed of an extraordinary skill in taking the fine views in a few minutes. For this purpose he made use of a machine which transmitted the landscape on a reduced scale to the paper. Between Valletri and Albano, where we went part of the way on foot in order better to enjoy the magnificent landscape and the mild air, we saw the whole method of his proceeding which afforded infinite pleasure to the children. He showed us afterward his collection of views, of which he had upward of two hundred of Naples and its neighbourhood alone. He gave me his address: Major Cockburn, Woolwich, nine miles from London.'[51]

In 1892 the *Journal* of the Royal Microscopical Society published the abstract of a paper by Dr Henry G Piffard which had appeared previously in a New York medical journal.[52] He compared the camera obscura and camera lucida as aids for making drawings through the microscope, although he found that the camera lucida required reasonably normal vision: 'The writer has for a long time, perhaps always, been affected with astigmatism and hypermetropia, to which advancing years have added presbyopia, and in consequence is unable to use the camera lucida with satisfaction.' He then described how he projected the microscope image using a right-angled prism with a silvered hypotenuse, and went on: 'The micrographic science of the future will seek the aid of the pencil less, and make more frequent use of the convenience and accuracy of photography. Bausch and Lomb made and mounted for me the prism described, and I have no doubt will be pleased to duplicate it for others.' The abstract of Piffard's paper also appeared in the *British Journal of Photography* on 18 August 1893.

W Jerome Harrison, in his book on *The History of Photography* (1887), declared Porta to be the inventor of the camera obscura.[53] Mr P C Duchochois, in a contribution to the American journal *Photographic Times* on 23 December 1887, declared that this was an error and although Mr Harrison's book had its merits, Porta had only perfected the camera obscura. Harrison was a staunch supporter of Porta as the inventor, and in the *British Journal of Photography* of 12 October 1888, he argued that Galileo and James Watt were known as inventors even though they had only 'perfected' the telescope and the steam engine. In the course of the argument he made the statement 'Inventing is like making love; no man can feel sure of having been "the first".' In subsequent issues of the *Journal* the 'Preface to the Reader' and Chapter XVII of Porta's *Natural Magic* were quoted by Harrison in order to prove his point. He concluded by saying 'as I see that splendid work, the American Encyclopaedia, declares him to be—the inventor of the camera obscura.'

One of the early issues of the *Photographic News*, a weekly magazine which started in 1858 and continued for fifty years (after which it became incorporated in the *Amateur Photographer*), contained an article entitled 'A catechism of photography' with the inevitable question and answer, 'By whom was the camera invented? By Giovanni Baptiste Porta a Neapolitan physician, about two centuries ago.'[54] In the following year John Spiller, a surgeon of the War Department, wrote a paper on 'The eye as a camera obscura'; he and Mr Allan B Dick succeeded in making a photograph using the eye of a bullock as the camera lens. This experiment was a photographic version of Kaspar Schott's demonstration in the seventeenth century.

In 1875 the magazine reported from an American journal: 'Among recent inventions is the application of the camera obscura to a railroad car, imparting to the travelling and wondering beholder a moving diminutive picture of the country through which he is passing.' In a 1901 issue was a letter containing a request for information on the optics required for making a camera obscura. The writer asked whether a lantern lens of eleven-inch focus would be suitable, and he requested construction details for the revolving top and wished to know if the summit of a hill would be a good site for the erection of the instrument. A satisfactory reply was given, but it would be interesting to know whether the correspondent, Mr G C Robinson, ever did build his camera obscura.

In 1974 the American journal *Horizon* reprinted a number of 'look into the future' items taken from various American magazines of the nineteenth century.[55] One of the drawings was of a 'Camera obscura for offices—much needed by business men', showing a man sitting in an office with his head through a small opening in a partition, looking at the table of a camera obscura. The image on the table showed several clerks in a general office lounging and reading newspapers. This drawing is of particular interest because of its similarity to the use of the modern closed-circuit television. It is also reminiscent of Joseph Harris who, in the eighteenth century, suggested viewing the screen of a

88. An illustration from an article on future developments published in an American journal of the nineteenth century.

"Camera obscura for offices—much needed by business men"

camera obscura through a small opening from an adjacent room which need not be darkened.

> *Why indeed*? Why does a photographer in the exercise of his business always use a black cloth? Why, of course, to make his camera obscurer. *Judy.*
>
> *British Journal of Photography* 22 November 1878

Appendix

The *Amateur Photographer* of 7 July 1899 quoted the *Art Journal* of 1861 on the making of a camera obscura out of a small packing case large enough to admit the head and shoulders.

While working on fluid lenses for telescopes, Peter Barlow, professor of mathematics at the Royal Military Academy, compiled his book *A New Mathematical and Philosophical Dictionary* (1814), which contained a brief history and description of the camera obscura. Barlow referred to the excellent camera obscura at the Royal Observatory, Greenwich, and 'a very good one made by Holroyd of Leeds, in which the images are received on a table more than six feet in diameter.' Several dictionaries mention Holroyd of Leeds, but a search has failed to reveal any further information about him or his camera obscura.

The *British Encyclopaedia* (1809) contained a brief description of two camera obscuras illustrated with clear diagrams, followed by a full discussion of Dr Wollaston's camera lucida.

Potonniée, in his *History of the Discovery of Photography*, referred to a M Cayeux who invented a camera obscura in 1809 'for the purpose of facilitating drawing by hand'.

A *Dictionary of Arts, Sciences and Manufactures* of 1842 stated that the camera obscura was 'an amusing optical instrument invented by the celebrated Baptista Porta.'

Encyclopaedia Londoniensis (1810) contained the usual account of Porta and described the construction of three different camera obscuras. Storer's 'Delineator' was mentioned, and the fact that the image 'will be found more sharp and vivid than those formed in the common instruments.'

An *Encyclopaedia of Photography* (1879) referred to Guyot's camera obscura, a concave surface for the table of a camera obscura and a prism with lens surfaces.

Jamieson's Dictionary (1827) described permanent and portable camera obscuras and added, in a note on perspective, 'a darkened room with a hole the size of a pea in a shutter, a white screen before the hole will receive an image in accurate perspective.'

Physikalisches Lexikon by O Marbach and C S Cornelius (Leipzig, 1850) contained an account of the camera obscura with many beautifully executed illustrations in fine white lines on black. The diagrams showed various types of camera obscuras and systems to make the image upright.

Two camera obscuras were mentioned in the *Catalogue of Optical, Mathematical, Philosophical and Chemical Instruments* by Watkins and Hill dated 1838. Portable pyramid instruments were priced between £4 14s 6d and £8 8s. The catalogue said 'This form of instrument has the advantage of the reflected image being viewed at once upon the paper intended for the drawing. No inversion takes place in the image, and a series of lenses are supplied for objects of different distances. By slight additions the Camera may be converted into an instrument for magnifying prints and drawings.' The second item was a simple box camera obscura at prices ranging from 14s to £3 3s 0d. Some twenty pages of the catalogue described a great variety of apparatus, microscope specimens and dissections, Claude Lorraine glasses and phantasmagoria lanterns.

The *Tunbridge Wells Visitor* of 20 September 1834 contained a news item about a camera obscura which gave 'panoramic views from our furze-clad common.' An advertisement in the same issue referred to the camera obscura as 'entirely new', that it was on the common for the season and it would be removed to different situations to provide a change of picturesque views. The admission charge was 6d per person. In praise of the image the advertisement said '. . . that any person may be distinctly recognised at a considerable distance.' This is an interesting comment when we read in Boswell's recent autobiography that his father's camera obscura was sometimes used to search for a lost child. The advertisement continued with a vivid description of the picture and ended with 'The focus of the camera obscura has been considerably improved . . .' This may mean that there had been a previous camera obscura at Tunbridge Wells.

The museum at Swansea, South Wales, has an undated photograph showing the sea front, railway station and pier at Mumbles. The pier was opened 10 May 1898 and so the photograph was probably taken in about 1900. A camera obscura can be seen on the sea front near the entrance to the pier, and appears to be a simple hexagonal timber structure. On the roof can be read 'Camera Obscura, Beautiful Scenery, Real Living Pictures, Only One Penny' and on the side 'Camera Obscura'. Similar photographs were reproduced in the following publications, each taken from a different viewpoint.

T.U.C. Souvenir Guide to Swansea 1901

89. A camera obscura at the entrance to Mumbles pier. (From an undated photograph at Swansea Museum.)

Swansea Improvements & Chamber of Trade Association 1903, 1905

South Wales *Daily Post* 1906, *Excursion Guide, Swansea and Neighbourhood*

The Swansea Library has a postcard (post date 1909) of the same scene without a camera obscura.

Douglas, Isle of Man: John Richard Fielding is referred to as a 'machinist' in a patent application made by him for 'improvements in camera obscuras' in November 1891. His address was Roach House, 2 Grosvenor Road, Douglas, Isle of Man. The patent no. 19,597 which was accepted in August 1892 describes almost exactly the camera obscura now to be found in Douglas.

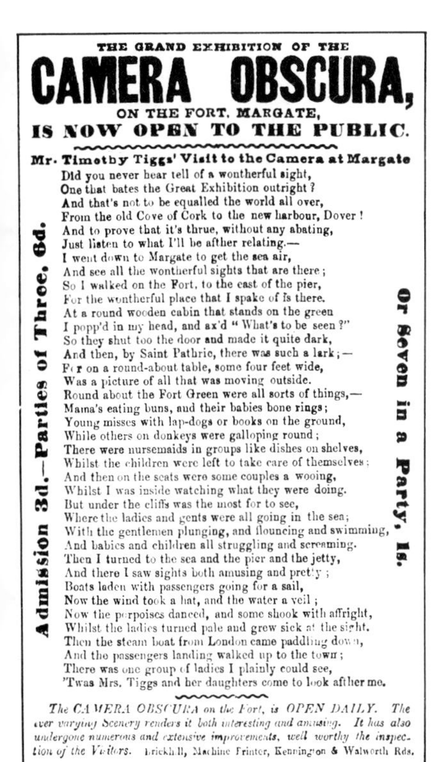

THE GRAND EXHIBITION OF THE

CAMERA OBSCURA,

ON THE FORT, MARGATE,

IS NOW OPEN TO THE PUBLIC.

Admission 3d.—Parties of Three, 6d.

Or Seven in a Party, 1s.

Mr. Timothy Tiggs' Visit to the Camera at Margate

Did you never hear tell of a wontherful sight,
One that bates the Great Exhibition outright?
And that's not to be equalled the world all over,
From the old Cove of Cork to the new harbour, Dover!
And to prove that it's thrue, without any abating,
Just listen to what I'll be afther relating.—
I went down to Margate to get the sea air,
And see all the wontherful sights that are there;
So I walked on the Fort, to the east of the pier,
For the wontherful place that I spake of is there.
At a round wooden cabin that stands on the green
I popp'd in my head, and ax'd "What's to be seen?"
So they shut too the door and made it quite dark,
And then, by Saint Pathric, there was such a lark;—
For on a round-about table, some four feet wide,
Was a picture of all that was moving outside.
Round about the Fort Green were all sorts of things,—
Mama's eating buns, and their babies bone rings;
Young misses with lap-dogs or books on the ground,
While others on donkeys were galloping round;
There were nursemaids in groups like dishes on shelves,
Whilst the children were left to take care of themselves;
And then on the seats were some couples a wooing,
Whilst I was inside watching what they were doing.
But under the cliffs was the most for to see,
Where the ladies and gents were all going in the sea;
With the gentlemen plunging, and flouncing and swimming,
And babies and children all struggling and screaming.
Then I turned to the sea and the pier and the jetty,
And there I saw sights both amusing and pretty;
Boats laden with passengers going for a sail,
Now the wind took a hat, and the water a veil;
Now the porpoises danced, and some shook with affright,
Whilst the ladies turned pale and grew sick at the sight.
Then the steam boat from London came paddling down,
And the passengers landing walked up to the town;
There was one group of ladies I plainly could see,
'Twas Mrs. Tiggs and her daughters come to look afther me.

The CAMERA OBSCURA on the Fort, is OPEN DAILY. The ever varying Scenery renders it both interesting and amusing. It has also undergone numerous and extensive improvements, well worthy the inspection of the Visitors. Brickhill, Machine Printer, Kennington & Walworth Rds.

90. Handbill advertising the camera obscura on the Fort at Margate during the 1840s and 1850s.

Chester: A keep on the city wall of Chester, known as Bonwaldesthorne's Tower, once housed a camera obscura, although nothing now remains of the instrument except an indication of a turret housing at the apex of the roof. The camera obscura is mentioned as a sight-seeing attraction in *The Stranger's Handbook to Chester* by Thomas Hughes (1856), and *Gresty and Burghall's Chester Guide* (1867). Nothing has been discovered of the origin or demise of this camera obscura.

Margate: The *Visitors New Guide to the Isle of Thanet* (1848) and *The New Historical Guide to the Isle of Thanet* (1848) both stated that 'On the Fort, opposite the Britannia Inn, stands the Camera Obscura.' An undated handbill announced the opening of the camera obscura on the Fort after it had undergone 'numerous and extensive improvements'. The bill consisted of a long humorous verse which refers to the camera obscura as a 'round wooden cabin' with 'a round-about table, some four feet wide', and it also mentioned the Great Exhibition of 1851. The library at Margate has a number of photographs and postcards showing a camera obscura on the sea front near the jetty, some of which bear posting dates from 1902 to 1911. The camera obscura was a temporary octagonal structure, showing various wordings and admission charges, and on one print it

91. The camera obscura on the jetty at Margate which operated from the late 1890s until 1910 or so.

is just possible to discern 'Read's camera obscura'. A search through engravings, prints and some directories, however, has brought no further information, but the library does have a large collection of unsorted local material.

The *Philosophical Magazine* of October 1844 contained a paper by G S Cundell entitled 'On a combination of lenses for the photographic camera obscura.' He referred to Wollaston's use of a meniscus lens and suggested mounting two such lenses in opposition about a central aperture. The concave surfaces were to face inwards, about one-seventh of the focal length of a single lens apart. It was claimed that the combination lens presented a flatter field, gave some chromatic correction and increased the working aperture for photography. This type of lens was made by Steinheil in

1865, and Kodak used a similar lens in a photographic box camera in about 1935.

Crystal Palace, Sydenham: A card from the patent weighing chair at the Crystal Palace has printed on the reverse 'Every visitor should pay a visit to the Camera Obscura situated upstairs in the Exhibition Gallery'. The card was written for Miss Clara Mosbery who weighed 8 stone 12 lb on the 13 July 1882. Mr M A Gilbert of the Crystal Palace Foundation, an archaeological group, has a programme dated 1910 of the entertainments at the Crystal Palace in which the camera obscura is mentioned. It was situated near the south-east corner of the gallery, was open from 10 a.m., and the admission charge was 3d. It is thought that Negretti and Zambra set up the apparatus in about 1870. Programmes dating from 1920 make no mention of the camera obscura, so it must have been dismantled during World War I, when the Crystal Palace was a naval training school. Gilbert, in a letter to the author, referred to a camera obscura at Beulah Spa, Upper Norwood, near Sydenham. The Spa opened in 1831 and closed in 1856, probably due to competition from the Crystal Palace. The camera obscura at the Spa was destroyed by vandals in about 1860.

In *Annalen der Physik* (1842, vol. LVI p407), Baron Ernst von Leyser described a 'camera clara dioptrica', using two sets of lenses, in which the image was presented correctly orientated and bright to the corners. The first set of lenses made an image of the scene before the camera obscura, and the second set projected that image onto a ground-glass screen; below this was a large condensing lens which illuminated the edges of the image. The achromatic lenses used throughout were very costly, particularly the large condensing lens which could be obtained only from Paris. Leyser explained that by suitable adjustment of the two sets of lenses it was possible to make a reduced or enlarged image on the ground-glass screen without moving the camera obscura. One might consider that he had made a crude zoom lens system! Finally, Leyser compared the daguerrotype with his camera obscura and noted many advantages in favour of the latter.

Karl Marx and Frederick Engels in their treatise *The German Ideology* (1846) wrote: 'If in all ideology men and their circumstances appear upside-down as in a camera obscura, this phenomenon arises just as much from their historical life-process as the inversion of objects on the retina does from their physcial life-process.'

The following nineteenth-century references are those to which no further information was given in the source or could be discovered by the author elsewhere. The source is indicated in brackets.

Castellan *Camera Obscura* (Potonniée)
Holroyd of Leeds, maker or owner of a camera obscura (*Pantalogia, A New Mathematical and Philosophical Dictionary* and *Jamieson's Dictionary*)
McAlister & Co., New York *Catalogue* (Gage and Gage)
Queen & Co., USA *Catalogue* (Gage and Gage)
Soleil 1812, constructed *'Pronopiographe'* (Potonniée)

References

1. Gernsheim H 1969 *History of Photography* (London: Oxford University Press)
2. Gernsheim H and Gernsheim A 1956 *L J M Daguerre: A History of the Diorama and Daguerrotype* (London: Secker and Warburg)
3. Newhall B 1949 *History of Photography* (New York: MOMA)
4. Potonniée G 1936 *History of the Discovery of Photography* (New York: Tennant and Ward)
5. Talbot, W H Fox 1844 *Pencil of Nature* (London)
6. *Revue Maritime et Coloniale* 1869 vol. **25**
7. Routledge R 1886 *Discoveries and Inventions* (London)
8. *Kilmarnock Standard Annual* 1957
9. *Kilmarnock and Riccarton Post Office Directory* 1840
10. McKay A (nd) *History of Kilmarnock*
11. Dumfries Museum *The Camera Obscura* (pamphlet)
12. Robertson S and Lord S (nd) *Camera Obscura*

13. *Some Remarks on the Future of the Observatory of the Astronomical Institution of Edinburgh* 1846
14. Willox J 1856 *Edinburgh Tourist and Itinerary*
15. Barr and Stroud Ltd 1968 *Catalogue*
16. Williams-Ellis, Clough 1971 *Architect Errant* (London: Constable)
17. *Guide to the Outlook Tower* (nd)
18. Glasgow Library (letter to the author)
19. Stranraer Museum (letter to the author)
20. Hull Town Clerk (letter to the author)
21. Henderson E 1838 *Arithmetical Architecture of the Solar System* (Glasgow)
22. Liverpool *Daily Post* 24 August 1966
23. Llandudno *Advertiser* 28 August 1966
24. *Western Mail* 30 December 1968; 3 January 1969
25. Jones, Barbara 1974 *Follies and Grottoes* (London: Constable)
26. Bristol City Art Galley 1973 *The Bristol School of Artists*
27. Morton H V 1927 *In Search of England* (London: Methuen)
28. Latimer J 1887 *Annals of Bristol in the 19th Century*
29. Cluer A 1974 *Plymouth and Plymothians* (London: Lantern Books)
30. Ramsgate Library (letter to the author)
31. Brighton *Herald* 5 September 1807
32. *Minute Book of the Proprietors of Brighton Suspension Pier* 1824–
33. Baxter 1824 *Stranger in Brighton*
34. Bishop J G 1897 *Brighton Chain Pier in Memorium*
35. Boore T H 1822 *Brighton Annual Directory*
36. Coghlan F 1838 *Brighton and its Environs*; 1862 *The Beauties of Brighton*
37. Whittemore J 1825 *Historical and Topographical Picture of Brighton*
38. Arnott, Neil 1865 *Elements of Physics* (London: Longman)
39. Atkinson E 1870 *Elementary Treatise on Physics;* 1900 *Natural Philosophy* (London: Longman)
40. Lardner, Dionysius 1855 *Museum of Science and Art* vol. 8 (London)
41. Brewster, David 1831 *Treatise on Optics* (London)
42. Guillemin A 1872 *Forces of Nature* (London)
43. Biot J B 1821 *Préçis Elémentaire de Physique Expérimentale* (Paris)
44. Wollaston, W Hyde 1812 On a periscopic camera obscura and microscope *Philosophical Transactions of the Royal Society*
45. Chevalier, Charles 1829, 1833 *Notice sur l'Usage des Chambres Obscures et des Chambres Claires* (Paris)
46. Marion F 1868 *Wonders of Optics* (London: Samson, Low, Son and Marston)
47. Imison J 1808 *Elements of Science and Art* (London)
48. Libri G 1838 *Histoire des Mathématiques en Italie* vol. 4 (Paris; reprinted Bologna 1967)
49. Rees A 1819 *Cyclopaedia*

50. *Magazine of Science* vol. 1 1839, 1840; vols 3 & 4 1842; vol. 6 1845
51. Spohr, Louis 1878 *Autobiography* (London) trans. Reeves
52. Piffard, Dr H G 1892 The camera obscura versus the camera lucida *The Microscope* vol. 12
53. *British Journal of Photography* Oct./Nov. 1888
54. *Photographic News* September 1858; December 1859; July 1901
55. *Horizon* 1974 vol. 16 (3)

The Twentieth Century

The camera obscura has almost been forgotten in the twentieth century. Neglect, fire and town planning have brought about the loss of many of the buildings but, fortunately, the few that remain appear to be well cared for and are open to the public. Camera obscuras are no longer built for public entertainment, but there is a current trend among museums to install camera obscuras in sections dealing with vision and optical aids for drawing. The Castle Museum in Nottingham, for instance, has recently constructed two camera obscuras as demonstration exhibits. One is situated in a wide window seat and demonstrates the image-making properties of a small hole, while the other has been installed in a picture gallery and is so arranged that a visitor may trace the image of a friend.

Observatories

The Great Union Camera at Douglas, Isle of Man, is perhaps one of the most remarkable of the camera obscuras still in operation.[1,2] The shallow, conical roof has eleven dormer windows, each containing a mirror and lens projecting an image onto a table. The eleven tables, arranged as a circle in the centre of the room, are separated from each other by a partition. As one walks around the group of tables a complete panorama is presented of Douglas Head, the town, the east cliff and the sea. Unfortunately, very little is known of the early history of this camera obscura, although the *Manx Sun* of 28 May 1887 (Jubilee year) did contain a note about a toboggan slide on Douglas Head, 'and the proprietor of the slide is also erecting a camera obscura in close proximity.' On 29 October in the same year the newspaper also reported that 'a fire took place at the camera obscura erected on Douglas Head, near to the switch-back railway which resulted in the total destruction of the

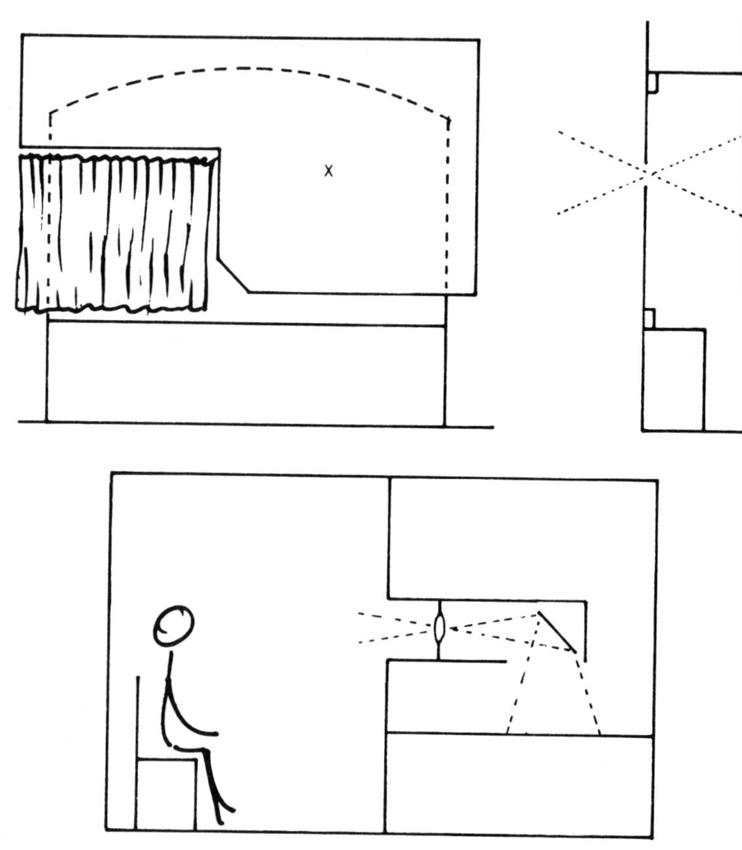

92. Two demonstration camera obscuras constructed in about 1975 at the Castle Museum, Nottingham. A simple one has been built in a window recess where the castle wall is about two feet thick. The window is covered by a board with a small hole (×) and another board, partly covering the recess, makes a screen which may be viewed when sitting behind the curtain. The size of the hole may be changed from about one inch to ¼ inch in diameter which demonstrates a sharper though dimmer picture. The second camera obscura consists of two cubicles, one open-ended for a sitter or model, and the other (the camera obscura itself) is enclosed except for a curtain on one side. Illumination of the sitter is sufficient for an image to be seen on the bench where it is possible to make a sketch. The lens is a single element of about 30 inches focal length, 2 inches in diameter.

building'. A strong wind from the south west contributed to the destruction of the camera obscura but it diverted the fire away from the switch-back and the slide. Optical instruments to the value of £100 were reported to have been lost in the blaze.

Robert Kelly wrote in *Manx Life* in 1976 that the camera obscura had been built on Douglas Head by James Fielding of Rochdale.[3] The site had been designated as recreational ground and consequently there was no official record of the building. Kelly wrote, 'Precisely when Mr. Fielding erected

93. A panorama camera obscura at Douglas, Isle of Man. The roof has eleven dormer windows, each with a lens and mirror projecting a view onto a table. The eleven tables are arranged in a circle in the centre of the room, and together they present an all-round picture of the town, the cliffs and the sea.

the Douglas Head camera and who he was remains a mystery.' However, Kelly referred to Porter's *Directory of the Isle of Man* of 1889 which mentioned the camera obscura and, although not connected, a hairdresser named James Fielding. Kelly continued the history with the sale of the building in 1907 to John Heaton, also from Rochdale, who had worked for Fielding at the camera obscura during the previous summer season.

The present owner of the camera obscura is Mr Norman Heaton, grandson of John Heaton, and son of Thomas Heaton (not grandson, as Kelly reports). Norman Heaton, in a conversation with the author, was under the impression that his grandfather had helped Fielding to build the camera obscura during the early years of this century (later than the reference in Porter's *Directory*) and that he had bought it from him in about 1905. Although uncertain about the past, Heaton thought that Fielding had been an engineer and an amateur scientist, and after describing the lenses in the camera obscura, he suggested that Fielding must have had a little knowledge of optics. The lenses in the Great Union Camera are of different focal lengths; the lens-to-subject distances are different, but the lens-to-table distances are all the same. Lenses giving views of the sea and the east cliff are exactly at their focal length away from the table because the view is at infinity, but those lenses showing parts of Douglas Head had to be of a slightly shorter focal length in order to give a sharp image of the closer view. Consequently, the lenses and turrets were numbered so that they can be easily replaced after cleaning.

This unusual camera obscura is in a very exposed position and therefore requires a great deal of care and maintenance. Should it ever become a burden to the owners it is to be hoped that the local authority will take it over. In complexity it stands between Dr Hoadley's eighteenth-century five-lens camera obscura at Chelsea, and the Russian 'Circorama', the complete 360° cinema shown in London during the 1960s. In this respect the Great Union Camera should be regarded as an historical monument.

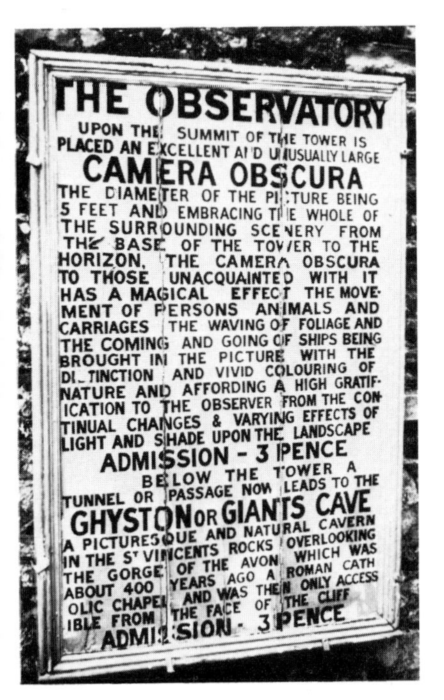

94. The notice board at the entrance to the Clifton Observatory at Bristol. The board is now (spring 1981) in a very dilapidated state.

In 1905 the Society of Merchant Venturers spent £550 on repairs to the Clifton Observatory in Bristol.[2,4] The commercial concession of the camera obscura was let out, and in 1929 an attempt by the lessee to increase business by installing 'automatic machines' was rejected by the Society. In 1866 William West's son wanted to set up a bazaar on the site but the Society did not approve. The Home Guard requisitioned the building in 1940, and it was not available to the public until after the war. The sale of the Observatory in 1977 to Honorbrook Inns Ltd was subject to conditions which included public right of access to the camera obscura. The Bristol architects Messrs Beecroft, Bidmead and Partners were commissioned to make plans for a complete renovation of the building to include a restaurant and a museum. The plans have been approved by the authorities but have yet to be implemented by the owners.

In about 1922 Sir Clough Williams-Ellis designed and built an Italian-style village at Portmeirion in North Wales, and during the early stages of construction he visited his friend and fellow architect Sir Patrick Geddes at the Outlook Tower in Edinburgh.[5,6] Sir Clough was fascinated by the camera obscura and decided to install one in his village. He designed a tower which was built at the end of the quay, right on the edge of the sea, and he asked Messrs Barr & Stroud to install the lens, mirror and table. Sir Clough, in a letter to the author, said that the parts used were second-hand, salvaged from a wartime submarine. The table is adjustable for focusing and the lens and mirror turret revolves to give a panorama of Portmeirion and the mountains of Snowdonia.

In 1925–26 Sir James Barrie, the playwright, donated a sports pavilion to the town of Kirriemuir and commissioned the architect Frank Thompson to design the building.[7] During the planning stage Barrie spent some time in Edinburgh and visited the Outlook Tower. He too was enchanted by the camera obscura and wrote to Thompson asking him to include one in the pavilion. This necessitated a redesigning of the building which, it is said, was a

95. The camera obscura at Portmeirion, North Wales. Sir Clough Williams-Ellis designed the tower and had it built at the end of the quay in about 1922. Photograph by courtesy of *Country Life.*

considerable improvement on the original drawing. The additional room for the camera obscura was octagonal, about twelve feet across, and when the building was completed in 1930 Messrs Negretti and Zambra supplied the camera obscura apparatus, which had an achromatic lens of 8 feet 9 inches focal length and an aperture of *f*15†. The mirror was originally at a fixed angle but it has since been made adjustable. The lens and mirror are housed in a rotating turret which is glazed for weather protection. The table is 4 feet 6 inches in diameter with a concave surface of about 6 feet radius, and focusing is achieved by moving the table up and down with a motorised jack to provide about 14 inches of movement. The sports field which is on high ground to the north east of Kirriemuir is an ideal situation for a camera obscura: by adjusting the mirror angle it is possible to view a cricket match in front of the pavilion or the Grampian peaks some fifty miles away.

In 1935 Horace E Dall, a fluid flow research engineer, designer and maker of microscope and telescope optics and amateur astronomer, built a camera obscura in the loft of his house at Luton in Bedfordshire. The lens, which he designed and made himself, is an apochromatic cemented doublet, with a diameter of 4¼ inches, a focal length of 135 inches, and a working aperture of *f*32. It is mounted with an optically flat, adjustable mirror in a revolving and retractable turret. The table is 24 inches in diameter, with a flat surface of matt white plastic, and has a motor-driven rise and fall movement of about 15 inches for focusing.

The image is extremely sharp, of high resolution and will withstand considerable magnification. Dall has also made a viewer‡ with interchangeable eyepieces for magnifications

†The *f* number refers to the aperture value of a lens and is determined by dividing the diameter of the lens or diaphragm into the focal length of the lens. In this instance the focal length is 105 inches and the diameter of the lens is 7 inches, making an aperture value of *f*15.

‡The viewing magnifier is similar to a focusing magnifier used with photographic enlargers.

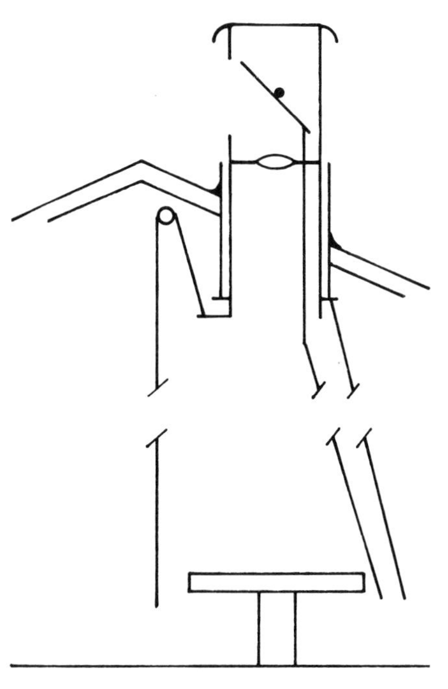

96. Diagram of the camera obscura built by Horace Dall in the loft of his house in 1935. The turret is raised and lowered by means of a rope; when in the closed position the cap fits over the outer fixed tube to make a weatherproof seal. The mirror angle and the rotation of the turret are adjusted by control rods which extend down to the table. The lens is 4¼ inches in diameter and about 12 feet above the table which is moved up and down to focus the image.

from 45× to 300×. An image of the sun is about 1¼ inches in diameter on the table, but when projected by means of the magnifier onto a wall of the loft it becomes 6 feet in diameter and is ideal for observing sunspots. This camera obscura was used to take a photograph of the transit of Mercury in 1960 which was reproduced in the book *Practical Amateur Astronomy* by Patrick Moore.[8] Dall has made many other photographs with his camera obscura—some for amusement, but usually in order to demonstrate optical resolution,

97. In about 1937 Dall used his camera obscura to take this snapshot of his neighbour at a distance of about 500 feet.

for which he has an especial interest. The combination of the magnifier and camera obscura lens is exactly the same as that in a projection microscope. In *Amateur Telescope Making: Advanced* Dall wrote:

> I projected in the dark (at night) a microscope slide of a flea, using the camera obscura like a projecting microscope, with a magnification of 500 on to a screen rigged up by a friend in a village over a mile away from here.[9]

Colonel Maude built a summerhouse type of camera obscura on his Bracken Hall estate at Shipley near Bradford, near to the pleasure gardens of Shipley Glen. This may have been one of the main tourist attractions of the Glen, with its swings and seesaws. The Reverend Robert Allan, writing in *The Dalesman,* remembered it as a hut large enough for about eight people; he said the 'table revolved', and that the entrance fee was one penny.[10] In a letter to the author Reverend Allan said the camera obscura was built around 1900 and demolished in about 1920. No further information has been gathered, although newspaper articles about the pleasure gardens suggest they were still open in 1955.

The pier at Eastbourne was built about 1901 under the direction of Mr Ridley, an architect, who also designed the superstructure consisting of entertainment rooms, tea-rooms and a camera obscura.[11] In 1970 a fire caused a great deal of damage to the building; the camera obscura was burned beyond repair and has not been replaced. Unfortunately no records have been found of the apparatus or the makers.

Arthur Gill, in a Royal Photographic Society publication (1976) wrote about a camera obscura at Grève de Lecq on the island of Jersey, which he remembered from about 1926 as being a circular, corrugated iron shed with a conical roof.[12] His article concluded with a remark on the possibility of the erection of a new camera obscura at Grève de Lecq but recent enquiries to St Hellier have provided no further information.

98. The camera obscura at San Francisco, one of the attractions of Cliff House on the southern shore overlooking Seal Rock. It was originally a timber structure erected in about 1939, but in 1951 it was rebuilt in concrete.

The camera obscura at San Francisco in California was erected in about 1939 by Mr F E Jennings and a magazine picture (source unidentified) now in the possession of Horace Dall shows that it was originally a timber construction. The present owner, Mr Gene Turtle, in a letter to the author, wrote that he was involved in the rebuilding of the camera obscura in 1951. The lens has a diameter of 6 inches, a focal length of 150 inches and a working aperture of about *f*25. It is mounted with an optically flat, aluminised mirror in a revolving turret. In order to maintain a good, clear picture the mirror has to be repolished every year because of corrosion by the sea air. The table is 6 feet in diameter and has a concave surface, which indicates that the lens may be a single-element biconvex lens. No further details are available except that the room is approximately twenty-five feet square. Mr Turtle said he was also involved in the construction of other camera obscuras at High Point, Colorado Springs, and at Rock City, Lookout Mountain, Chattanooga, Tennessee.

The camera obscura at Santa Monica, California, is now in a specially-built room at the City Senior Recreation Center.[13] It was originally built on the beach and was opened to the public on 21 April 1899, but was moved early in the 1900s to a more sheltered position in Palisades Park. The final move to the Recreation Center was in November 1955. The only technical information available is that the lens and mirror are housed in a revolving turret and that the image is received on a circular table.

A camera obscura costing about £1000 (autumn 1980) is one of the items listed in the catalogue of the Bedford Astronomical Supplies of Luton, England, a company founded in 1972 by Mr P R Drew, an engineer and amateur astronomer. The design of the camera obscura is similar to the one made by Dall, who is well known to Drew through a common interest in astronomy. Drew supplied his first camera obscura to Mr Norman Humphries of Haversham in Buckinghamshire, who installed it in a summer house and uses it to study the activity of wild fowl on nearby gravel pits. Humphries is able to project an image of the sun to a diameter of about 24 inches in order to observe sunspots. In January 1980 Drew made another camera obscura for Mr W W Mellor of Sheffield, an amateur astronomer who uses the instrument for solar and lunar observations and reports lunar occultations to the Royal Observatory at Herstmonceaux. Mellor recently published details of his observations of Venus and a note on his camera obscura in *The Inner Planets Newsletter* (the bulletin of the British Astronomical Association, Terrestrial Planets Section: No. 2, April 1980). Occasionally, Mellor also uses the camera obscura simply for the pleasure of looking at the intensified colouring of cloudscapes and sunsets. Drew is now designing, for his own personal use, a camera obscura with an increased aperture of 6¼ inches, focal length 110 inches and the table will be 30 inches in diameter.

Military Applications

During the 1920s the Harmsworth Publishing Organisation produced many encyclopaedias giving brief accounts of the camera obscura. The Harmsworth *Universal Encyclopaedia* of 1922 contains an illustration of a camera obscura which appears to be a prefabricated hut. The text reads: 'Apart from its lineal descendant, the photographic camera, the instrument survives only as a sideshow at holiday resorts, but it was used for observation purposes in the Great War.' Nothing further has been discovered of this application.

During the first world war the Royal Air Force used the large room-type camera obscura to observe and record the accuracy of bomb aiming during pilot training.[14] For this purpose a mirror is not required, since the lens is set in the ceiling of the room and projects a view of the sky. An *Air Publication* of 1917 gave instructions for making a camera obscura, which, although a simple building, 'must be fairly dark inside when the door is shut.' Very little was mentioned about the lens other than a method for finding the focal length and that it 'will be taken down and kept in a box when not in use.' The *Publication* went on: 'The camera obscura is designed for the following purposes,

I. to train pilots to fly over a target;
II. to train pilots to signal when exactly over a target;
III. to train pilots in bomb dropping.'

In order to calculate the speed of an aircraft it was necessary to know the focal length of the lens, the height of the aircraft (received by radio-telephone from the pilot) and the speed of the image of the aircraft. This was obtained by plotting the image on the table of a camera obscura and timing the measured plots with a metronome. The necessary formulae were given with working examples and the possible standard of accuracy to be achieved. Another *Publication* was issued in 1918 with a considerably shorter text. By 1924 the contents of the pamphlet on camera obscuras had increased to include an illustration of the lens curvature of field, which necessitated the use of a concave table surface for sharp focusing. When calculating the speed of aircraft, however, the camera obscura table had to be flat in order to measure the plotted courses of aircraft, and the instructions suggested that a compromise could be reached by focusing a sharp image between the centre and the edge of the table. The text emphasised the need for accuracy, and a method was given for finding the 'centre' of a lens. In 1932 an Order was issued by the Air Ministry on the standardisation of paper for use in camera obscuras, but added that 'Unit Commanders are to ensure that economy is observed in the use of the paper'! The camera obscura was still in use by the Royal Air Force for bombing practice during the second world war.

The Royal Australian Air Force also used the camera obscura during this period and issued a foolscap leaflet of lecture notes on its location, erection and uses.[15] The title page bore a vigorous and amusing sketch of a camera obscura tent bulging with bodies and betting phrases issuing forth. The notes mentioned that both permanent and portable camera obscuras were used in the service. The portable model consisted of a tripod holding both the lens and table, and a canvas cover with a hole fitting round the lens was draped over the tripod in order to make a light-proof tent. It was used

a) to test a pilot's ability to fly a straight, level and steady course.
b) to find wind velocity by day or night.
c) simulation of level bombing (day or night).

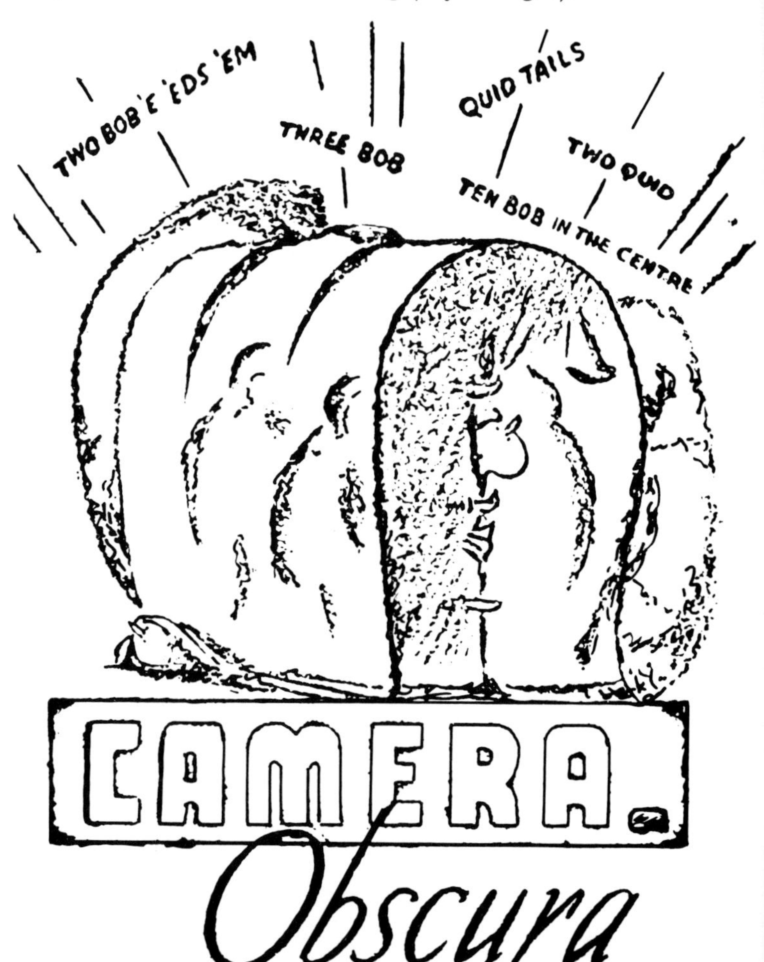

99. The cover design of lecture notes on the use of the camera obscura, issued by the Royal Australian Air Force in 1943. The lens is in the roof of the tent just below the words 'three bob'. An image of the sky and aircraft was made on the table in the tent, and from observations and measurements it was possible to calculate whether a simulated bomb drop was on target.

Full instructions and formulae were given, and a metronome was used for timing.

The Royal Aircraft Establishment at Cardington has a camera obscura for the measurement and analysis of wind movement of clouds and the oscillations of tethered balloons. The Establishment at Orfordness erected a permanent camera obscura around 1928–30 for the accurate determination of wind direction and velocity at various altitudes. This instrument may also have been used to locate and measure precisely the splash marks of practice bombs dropped into the sea.

Art and Drawing

The 'Copy-Scanner' made by Halco-Sunbury Co. Ltd is a modern camera obscura in which the function and basic design are recognisable as developments from the camera obscuras of Newton (1844) and Bion (1727). The 'Copy-Scanner' is used by graphic artists for redrawing a design to a larger or smaller scale, allowing the artist to see the effect of a change of scale before drawing. Although a similar result can be achieved by photography the camera obscura method is much quicker, and it also retains continuity of the creative process.

The Edmund Scientific Company of Barrington, New Jersey, USA, specialises in marketing new and surplus apparatus, and in 1976 one of their mail-order catalogues

100. The 'Copy-Scanner' made by Halco-Sunbury Ltd for copying pictures, flat objects and transparent material. The ground-glass screen has a large hood to shield off extraneous light and is at a fixed height from the floor. The lens and base board may be raised or lowered in order to focus and adjust the size of the image. The centre portion of the base board is a light box for transparent pictures or when silhouettes are required. The evolution of this apparatus may be seen through the camera obscuras of Bion (p73) and Newton (p130). A completely enclosed model of this copier, the 'CopiKam', is the latest development by Halco-Sunbury.

carried the following advertisement:

> Trace landscapes and portraits from life. Make a camer obscura, aim its lens at the scene you wish to draw, it appears i colourful detail on 11⅞″ sq. screen. Then trace a propor tionally-correct reversed picture. Kit contains 3⅜″ diam. 22½″ F.L. pcx. lens; tube; focussing sleeve; plastic Fresnel screen You need wood or cardboard and a mirror to complete. Instr N. 70,881 $17.75 Ppd.'

A recently made reflex box camera obscura, probably a replica, may be seen at the Fox Talbot Museum, Laycock Abbey. It rests on the sill of a deep-set window and a small platform two or three steps high enables visitors to view the image on the ground-glass screen which is about 8×6 inches. Among the museum exhibits there are several camera obscuras used by Fox Talbot for his experiments in the photographic process. One of these small instruments, which is about 7×4×2½ inches, is an interesting example of the craftsmanship of the period.

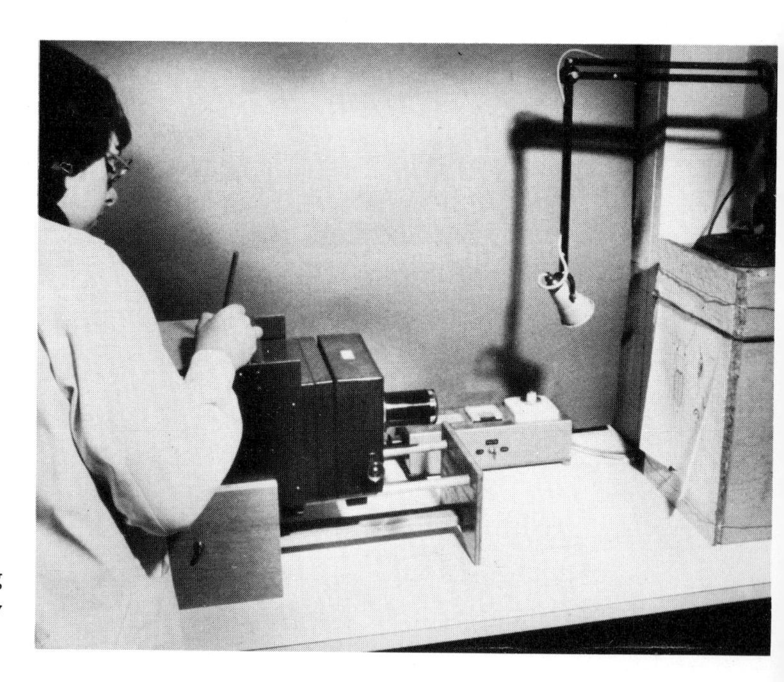

101. A camera obscura for copying drawings, made in a laboratory workshop.

Literature—Fiction

Vladimir Nabokov titled one of his novels *Kamera Obskura* when the original Russian text was published in Paris in 1933. The English text, first published in the USA in 1938, was given a new title *Laughter in the Dark*, and was subsequently reprinted in England in 1969. The story is of a wealthy art patron and critic infatuated with a cinema usherette who becomes his mistress. They have a car accident in which he is blinded and she takes advantage of this by entertaining a previous lover in the house. The blind man suspects the truth and tries to shoot her, but stumbles and shoots himself. The title *Kamera Obskura* is echoed by Nabokov's use of the cinema, blindness and the darkness of death.

One of Alfred Hitchcock's anthologies of short stories includes a contribution from Basil Copper also entitled *Camera Obscura*, which tells of a money-lender who calls on a client for a repayment.[16] He is invited into a room which is a camera obscura and shows a view of the town. Later he is taken to another room where a special camera obscura shows the town as it had been some years previously; a scene of poverty and degredation. The money-lender is finally shown out of the house but by a different door from that which he entered. He discovers he is in the past of the camera obscura picture and he meets other money-lenders. The streets are unfamiliar, and he tries to find his way groping unsuccessfully—he has entered death.

The Reverend R H Malden, Dean of Wells, wrote an engaging series of short stories under the title *Nine Ghosts*. A passage from *The Coxswain of the Lifeboat* reads:

> It was exactly like what one used to see in a camera obscura when I was a child. Some ingenious arrangement of lenses, and mirrors too I suppose, threw a picture of what was passing onto a table in a darkened room. I suppose such a thing hardly exists now. It could not hope to compete with the films; though as a matter of fact I came upon one only ten years ago in a queer old house in Edinburgh, not far from the Castle.

This description is used as an analogy for the sensation of a dream, although the narrator is convinced he did not fall asleep! John Moore in his novel *Brensham Village* refers to a folly built in 1755 on the 'Hill'. The folly was owned by a lordly hermit and contained a non-working camera obscura and a telescope with a broken lens.

H G Wells, the novelist and historian, was particularly influenced by science and technology. In one of his short stories *Land of the Ironclads* (1903), he conceived a war machine similar to a tank. He described an armoured vehicle with rows of rifles which could be operated by pressing a button, 'and they had the most remarkable sights imaginable, sights which threw a bright little camera-obscura picture into the light-tight box in which the rifle man sat below. This camera obscura picture was marked with two crossed lines, and whatever was covered by the intersection of these two lines, that the rifle hit. The rifleman stood up in his pitch dark chamber and watched the little picture before him . . . As he pushed the knob about the rifle above swung to correspond, and the pictures passed to and fro like an agitated panorama.' This literary refinement of the experiment at Anvers in the 1860s and the adaptation of Leonardo's drawings of armoured vehicles is typical of Wells' style, his research and inventive vision.

The film *A Matter of Life and Death* (American title *Stairway to Heaven*), made by Archer Productions in 1946, featured a camera obscura for the first time in a motion picture. The stars were Roger Livesey, David Niven and Kim Hunter, directed by Michael Powell and Emeric Pressburger. In a letter to the author, Mr Powell said that the camera obscura at Portmeirion gave him the idea of using one in the film. One of the characters, Dr Reeves (Roger Livesey) referred to the camera obscura as being like 'God's eye' which he used to observe the life and activity of the village where he lived, and he often arrived at diagnoses from what he saw. The camera obscura depicted was of the conventional summerhouse type. The picture of the village on the table was simulated by the camera men, Jack Cardiff and Geoff Unsworth, who shot both scenes separately and combined them by optical processing.

Literature—Non-fiction

Joyce Hemlow in her book *History of Fanny Burney* used the term 'camera obscura' when describing Fanny. Early in the biography she wrote: 'Now Fanny was sixteen and she observed the life around her, listened and reflected, and no one suspected as yet that she had a memory like a tape recorder, a mind like a camera obscura.' Later when Fanny was in France with her husband, 'The public, remembering the comic realistic scenes of Evelina and Cecilia, had expected Madame D'Arblay to focus a camera obscura behind the Napoleonic curtain.'

H V Morton wrote *In Search of England* in 1927 and mentioned the camera obscura at Bristol. He described the scenes he saw at 'each turn of the table', and continued 'As I stood there in the dark I thought that a few hundred years ago any man who owned a camera obscura would have been burned at the stake or made Lord Chancellor.' Morton concluded his account of the Clifton Observatory with 'Why, you wonder, has no writer of detective stories used the camera obscura? The eye of the hill lends itself to treatment.'

Major-General J Waterhouse gave a lengthy and informative lecture on the history of the camera obscura to the Royal Photographic Society in 1901, starting the history with Aristotle's observation and ending with the invention of photography. [17] The appendix to the published paper contained many extracts from original or early editions of texts on optics and the camera obscura.

A recent booklet by Scott Robertson and Simon Lord described and illustrated the camera obscura and its history.[2] The four working camera obscuras at Edinburgh, Douglas, Dumfries and Bristol are also illustrated and described with notes on their history.

The *British Journal of Photography* has continued to publish articles on the camera obscura, but as late as 1901 the term was still used to describe a stereoscopic photographic camera. H Rawlinson contributed a description and historical account titled 'Camera obscura' in 1953, and Dr Ollerenshaw referred to Cheselden in an article on the use of the camera obscura in medical illustration.[18,19] The present author made a contribution to the *Journal* in 1976 dealing with the construction and use of a camera obscura for redrawing a variety of illustrations of insects to a uniform size.[20,21] In another note published in 1977, the name of the instrument was discussed and an attempt made to justify the use of the term 'camera obscura' as an English name, with the plural 'camera obscuras', in place of the original Latin. Other articles in the *Journal* have been mentioned in the chronology of the subject matter.

Dictionaries and encyclopaedias usually contain articles on the camera obscura, but the texts are often brief and inadequate. However, the article in the eleventh edition of the *Encyclopaedia Britannica* (1910) is probably the most comprehensive, with a full account of the development of the instrument. Referring to the periscope, *Britannica* stated that it was a 'camera obscura under a new name'. *Everyman's Encyclopaedia* (1967) added to the occasional outbreaks of controversy about the invention of photography by saying 'the camera obscura was first employed in the interests of photography about 1794 by Thomas Wedgwood.' The *Dictionary of Art Terms and Techniques* (1969) by Ralph Mayer gives a satisfactory definition of a camera obscura and ends with an extraordinary statement: 'For practical purposes the portable camera obscura has long been supplanted by the camera lucida and by the 20th century large screen developments.'

Conclusion

The portable camera obscuras of the eighteenth and nineteenth centuries have now become collector's items, and the amateur artist has taken instead to that 'lineal descendant', the photographic camera. As an aid to drawing the camera obscura remains recognisable in its original form only in commercial art studios. In science and technology, lens projection systems bear little resemblance to the camera obscura, but on rare occasions a need arises for the more simple instrument. However, the large structural camera obscura remains unchanged and may even regain some of its original attraction as a popular sideshow. As an adjunct to tourism it has the charm of antiquity and the useful function of displaying a panorama of a town or countryside.

> The prettiest landskip I ever saw was one drawn on the walls of a dark room.
>
> Addison (1712)

Appendix

Notes from a conversation with Mr Horace Dall, 2 March 1979.

A tent camera obscura was seen by Dall on the beach at Clacton-on-Sea in 1917.

A camera obscura at Blackpool during the 1920s was a War Department surplus item.

A camera obscura was built by Dall for Colonel Muirehouse in Cornwall about 1946; the Colonel died in about 1950 and the camera obscura was moved to Falmouth.

In 1977 Dall made the optics for a camera obscura for Mr Bill Southern at Polperro in Cornwall. It is uncertain whether it was completed because Southern was very ill at the time.

Dall constructed the optics for camera obscuras which were erected at Alton Towers, Cheshire, and Blackpool for Mr K Dutton of Workington. The Alton Towers instrument was used for public viewing for some years. [The present author wrote to Dutton in 1975 but the envelope was returned marked 'unknown'.]

Mr F E Jennings of San Francisco visited Dall in 1937–38. As a result he built the San Francisco camera obscura in about 1939

> and sent Dall a photograph of Seal Rock, which he still has, taken with the camera obscura soon after it was erected. Jennings may have been responsible for the original buildings of the camera obscuras at High Point, Colorado, and Rock City, Tennessee.

The *Evening Standard* of 9 March 1979 reported that the Romm Doulton Organisation, a business group based in California, proposed to convert the Royal Agricultural Hall at Islington (North London) into a Charles Dickens fantasy land, and among the many attractions would be a camera obscura.

S G Boswell, in his autobiography *The Book of Boswell* (1970), refers to a camera obscura managed by his father on the south shore at Blackpool early in the century. It was hexagonal, probably lightly constructed, and used for entertainment and, on occasion, to search for a lost child.

In 1944 Cecil Bathurst of Wolverhampton applied for a patent entitled *Improvements in Camera Obscuras* '... which will afford an amusement and educational appeal to young persons and which can be produced at relatively low cost.' The design was a miniature room-type camera obscura with a rotating turret and an eyepiece in the side of the 'room'. The patent number 591,785 was granted in August 1947, and a summary was published in the *British Journal of Photography* on 10 June 1949.

At an auction sale held by Christie's South Kensington Ltd on 21 August 1980 a late eighteenth-century camera obscura was sold for £1800. It was a folding model, 10 inches long, with a scioptric ball lens, a focusing screen 3¼ × 3¾ inches, and an ivory plate engraved 'Clover Professor of Optics to His Majesty No. 169', although this name has not been traced among instrument makers. Another camera obscura (early nineteenth century) made of mahogany, 11¾ inches long and with a focusing screen 5×6 inches, was sold for £750. In this model focusing was achieved by a sliding box and the lens was in a mahogany mount with a screw-fitting cap. A third camera obscura of about 1880 made in walnut, also had a sliding-box focusing arrangement fitted with a brass-mounted lens. The instrument was without a focusing screen, and was sold for £200.

At an auction held by Messrs Sotheby, Belgravia, on 20 March 1981 a mahogany, box focusing, reflex camera obscura was sold for £1400. It was about 7 inches long with a ground-glass screen 2¾ inches square, a single-element lens with a mahogany screw-fitting lens cap. The camera obscura was stated to be early nineteenth century.

102. A late eighteenth-century folding camera obscura by Clover, sold by Christie's South Kensington Ltd in August 1980 for £1800. The lens is mounted in a scioptric ball, an unusual feature for a portable camera obscura.

An amusing article by Philip Hypher titled 'Obscure camera' appeared in the *British Journal of Photography* on 27 March 1981. Although humorous it contained long passages paraphrased from Battista Porta's *Magiae Naturalis.*

The terms camera lucida and camera obscura continue to be confused. Recent research has discovered many camera obscuras and associated projection apparatus which have been classified as camera lucidas in the patent literature of the nineteenth and twentieth centuries.

References

1. *Manx Sun* 28 May 1887; 29 October 1887
2. Robertson, Scott and Lord, Simon 1978 *Camera Obscura* (Outlook Tower, Edinburgh)
3. *Manx Life* November/December 1976
4. Robertson, Scott 1959 *Notes on the Clifton Observatory* (pamphlet
5. Williams-Ellis, Clough 1971 *Architect Errant* (London: Constable
6. *Guardian Weekly* 16 April 1978
7. Kirriemuir Local Authority *Notes on the Camera Obscura Fitted in the Sports Pavilion*
8. Moore, Patrick 1963 *Practical Amateur Astronomy* (Guildford Lutterworth)
9. Ingalls A G 1963 *Amateur Telescope Making: Advanced* (Scientific American)
10. *The Dalesman* September 1978
11. Eastbourne Pier Company (letter to the author)
12. Gill A J 1976 *Camera Obscura* (Royal Photographic Society Historical Group)
13. Santa Monica City Administration *What is a Camera Obscura?*
14. Royal Air Force AML 1674 1925; AMOs A8-A14/ 1932; *Air Publication* 242-1917, 356-1918, 961-1924
15. Royal Australian Air Force 1943 *Publication* 296
16. Hitchcock, Alfred 1967 *Camera Obscura,* in *Stories That Scared Even Me* (London: Reinhardt)
17. Waterhouse J 1901 *Photographic Journal*
18. *British Journal of Photography* April 1953
19. *British Journal of Photography* September 1977
20. *British Journal of Photography* April 1976
21. *British Journal of Photography* July 1977

Bibliography

*Indicates not seen by the author

*Addison, Joseph 1712 *The Spectator* 25 June
*Algarotti F 1763 *Saggio Sopra la Pittura* (Levorno)
—— 1764 *Essay on Painting* (Glasgow)
*Alhazen *Opticae Thesaurus* (Basileæ)
Allan R C 1978 Down in the glen *The Dalesman* September
Allemand J N S 1774 *Œuvres Philosophiques et Mathématiques* (Amsterdam)
Alston R W 1954 *The Painter's Idiom* (London: Staples)
**Amateur Photographer* 7 July 1899
American Cinematographer July 1947
Annalen der Physik 1842 vol. LVI p407
*Archer Productions Ltd (film company) 1946 *A Matter of Life and Death*
Aristotle (*see* Hett W S)
Arnott, Neil 1865 *Elements of Physics* 6th edn (London: Longman)
The Artist January 1980
Arts Council 1972 *From Today Painting is Dead* (London: HMSO)
**The Athenaeum* 22 February 1845, p202
Atkinson E 1870 *Elementary Treatise on Physics* trans. D Ganot
—— 1900 *Natural Philosophy* (London: Longman)
Aubrey, John *Brief Lives* 1978 ed O L Dick (Harmondsworth: Penguin)

*Bacon, Francis 1623 *De Augmentis Scientorum*
*Bacon, Roger *Perspectiva* (Frankfurt)
—— *De Multiplicatione Specierum*
*Barbaro D 1568 *La Pratica della Perspettiva* (Venice)
Barlow P 1814 *New Mathematical and Philosophical Dictionary* (London)
Barocas V (*see* Ronchi V)
Barr & Stroud Ltd 1968 *Catalogue*
Baxter 1824 *Stranger in Brighton*
Belchier, John 1773 An account of a book entitled *Osteographia,* in *Philosophical Transactions of the Royal Society No.* 430 p194
*Benedetti G B 1585 *Diversarum Speculationum Mathematicarum et Physicarum Liber* (Turin)
*Benjamin Park *Intellectual Rise in Electricity*
Bettinus, Marius 1645 *Apaira* (Bononiæ)

Bindman D and Puppi L 1970 *Canaletto, the Complete Paintings* (London: Weidenfeld and Nicolson)
*Bion, Nicolai 1727 *Neu-eroffrete Mathematische Werke-Schule . . .* (Nuremberg)
—— 1723, 1728 *Construction and Principal Uses of Mathematical Instruments . . .* transl. Edmund Stone (London)
Biot J B 1821 *Précis Elémentaire de Physique Expérimentale* (Paris)
Bishop J G 1897 *Brighton Chain Pier in Memorium* (Brighton *Herald*)
Blunt R 1923 *Mrs. Montague, Queen of the Blues, Letters and Friendships from 1762–1800* 2 vols (London: Constable)
Book of British Towns 1979 (London: AA, Drive Publications)
Boore T H 1822 *Brighton Annual Directory*
Boswell S G 1970 *The Book of Boswell* (London: Gollancz)
Boyle R 1669 *Of the Systematical and Cosmical Qualities of Things* (Oxford: Royal Society)
Bradbury S and Turner G L'E 1967 *Historical Aspects of Microscopy* (Cambridge: Heffer)
Bragg W H 1933 *The Universe of Light* (London: Bell)
Braikenridge Collection at the Central Library, College Green, Bristol, vol. 29, pp303, 309, 314, 319, 325
Brander G F 1764 *Polymetroscopium Dioptricum oder Beschreibung eines Optischen Instrumentes—die Bilder in einer Camera Obscura* (Augsburg)
*—— 1766 *Beschreibung einer Camera Obscura und Sonnen Mikroskops* (Augsburg)
*—— 1769 *Beschreibung dreyer Camerae Obscurae* (Augsburg)
—— 1769 *Kurze Beschreibung einer ganz neuen art einer Camerae Obscurae* (Augsburg)
—— 1775 *Kurze Beschreibung der neu Abgeanderten—Camera Obscura* (Augsburg)
Brewster D 1831 *Treatise on Optics* (London)
—— 1805 *Ferguson's Lectures on Select Subjects* (Edinburgh)
Brighton
Brighton Herald 5 September 1807
Minute Book of the Proprietors of Brighton Suspension Pier (1824)
Pinnock's Guide to Knowledge (31 March 1835)
Baxter's Stranger in Brighton (1824)
Bishop J G 1897 *Brighton Chain Pier in Memorium*
Boore T H 1822 *Brighton Annual Directory*
Coghlan F 1838 *Brighton and its Environs*
—— 1862 *Beauties of Brighton*
Whittemore J 1825 *Historical and Topographical Picture of Brighton*
Bristol
Buchanan R A and Cossons N 1970 *Industrial History in Pictures: Bristol* (Newton Abbot: David and Charles)

Morton H V 1927 *In Search of England*
Latimer J 1887 *Annals of Bristol in the Nineteenth Century*
British Journal of Photography Oct, Nov 1888; March 1889; May 1891; Aug 1893; April 1901; May 1901; Dec 1916; Nov 1916; April 1953; April 1976; July 1977; Sept 1977; March 1981
Brown T 1901 *British Journal of Photography* 5 April
Bruce, James 1790 *Travels to Discover the Source of the Nile* (Edinburgh)
Buchanan R A and Cossons N 1970 *Industrial History in Pictures: Bristol* (Newton Abbot: David and Charles)
Burke, J 1955 *William Hogarth, 'The Analysis of Beauty' with Rejected Passages* (London: Oxford University Press)
*Busch G 1775 *Encyclopadie der Historischen, Philosophischen und Mathematischen* (Hamburg)

Camera lucida (*see* Fyffe G)
*Cappi A 1839 *Delle Applicazione dell'Italiano Cellio e del Francese Daguerre alla Camera-obscura*
*Cardano, Girolamo 1550 *De subtilate* libri XXI (Nuremburg)
Carter B A R 1962 *Review, Burlington Magazine* September
Cato J 1955 *Story of the Camera in Australia* (Melbourne)
*Cellio M A 1686 *Descrizione d'un Nouvo Modo di Trasportar . . .* (Rome)
Ceram C W 1965 *Archaeology of the Cinema* (London: Thames and Hudson)
*Cesare Cesariano 1521 *De Architectura Libri dece Traducti de Latino in Vulgare Affigurati* (Como)
Chenakal V L 1953 *Russian Tool and Instrument Makers of the Eighteenth Century* (trans. Smithsonian Institution, Washington 1976)
*Cherubin d'Orleans 1671 *La Dioptrique Oculaire* (Paris)
Cheselden W 1773 *Osteographia* (London)
Chevalier, Charles 1829, 1833 *Notice sur l'Usage des Chambres Obscures et des Chambres Claires* (Paris)
Chevalier V *Avis aux Amateurs des Beaux-arts . . .* (Paris)
Clark, Kenneth 1949 *Landscape into Art* (London: Murray)
Claudet A 1857 On the phenomena of relief of the image . . . *Proceedings of the Royal Society* vol. 8
Clay R S and Court T H 1932 *History of the Microscope* (London)
*Clerc, Alexis 1884–87 *Physique et Chimie Populaires* (Paris)
Clerc L P 1947 *Photography, Theory and Practice* (London: Pitman/Greenwood)
Cluer, Andrew 1974 *Plymouth and Plymothians* (London: Lantern Books)
Coe, Brian 1977 *Birth of Photography* (London: Ash and Grant)
Coghlan F 1838 *Brighton and its Environs*
—— 1862 *The Beauties of Brighton*

Cohen B I 1950 *Some Early Tools of American Science* (Cambridge, Mass: Harvard University Press)
Coke, Van Deren 1975 *One Hundred Years of Photographic History* (Albuquerque: University of New Mexico Press)
Collins-Baker C H 1921 *Crome* (London: Methuen)
Comptes Rendus Académie des Sciences (Paris) 1841 vol. 13 p234
Constable W G 1962 *Canaletto* (London: Oxford University Press)
Cooke, Olive 1963 *Movement in Two Dimensions* (London: Hutchinson)
Copper, Basil 1967 *Camera Obscura* In *Stories That Scared Even Me* ed Alfred Hitchcock (London: Reinhardt)
⋆*Court Miscellany* April 1770
⋆Croker's *Theological Dictionary* (London)
—— 1773 *Complete Dictionary of Arts and Sciences*
Crommelin C A 1951 *Catalogue of the Physical Instruments of the Eighteenth Century in the Rijksmuseum* (Leyden)
Cuff, John (published for) 1747 *Verses Occasioned by the Sight of a Chamera Obscura*
Curz M (*see* Levi Ben Gershon)

The Dalesman September 1978
Dall H E (*see* Ingalls A G)
⋆Danti, Ignatio 1573 *La Prospettiva di Euclide* (Florence)
Darwin, Erasmus 1789, 1791 *The Botanic Garden* (London)
⋆Derham W 1726 *Philosophical Experiments and Observations of the Late Eminent Dr. Hooke* (London)
⋆Descartes, René 1637 *La Dioptrique* (Paris)
⋆Deschales C F M 1674 *Cursus sen Mundus Mathematicus*
Dickes W F 1905 *The Norwich School of Painting* (London and Norwich)
Digges L 1571 *Pantometria* (London)
⋆Dodwell, Edward 1819 *A Classical and Topographical Tour Through Greece* vol. 1 (London)
Duhem P 1954 *Le System du Monde* vols 3, 4 and 5 (reprint Paris)
Dumfries Museum (nd) *The Camera Obscura and What You See* (pamphlet)

⋆Eder J M 1945 *Ausfuhrliches Handbuch* vol. 1
—— 1945 *History of Photography* (New York: Columbia University Press)
Edgerton S Y 1975 *Renaissance Rediscovery of Linear Perspective* (New York)
Edinburgh
Some Remarks on the Future of the Observatory of the Astronomical Institution of Edinburgh (pamphlet, 1846)
Willox J 1856 *Edinburgh Tourist and Itinerary*
Barr & Stroud Ltd 1968 *Catalogue*

Williams-Ellis, Clough 1971 *Architect Errant* (London: Constable)
Guide to the Outlook Tower
Edmund Scientific Corp, USA 1976 *Catalogue*
Encyclopaedias and Dictionaries
British Encyclopaedia 1809
Cassell's Cyclopaedia of Photography 1911
Century Dictionary 1889 (ed W D Whitney) (New York)
Chambers Cyclopaedia 1741, 1860, 1888, 1973
Croker's Dictionary of Arts and Sciences 1773
Croker's Theological Dictionary
Dictionary of Art and Artists 1965
Dictionary of Arts, Sciences and Manufactures 1842
Dictionary of Art Terms and Techniques 1969
Dictionary of Photography 11th edn
Dictionary, New Complete, of Arts and Sciences 1778
Dictionaire Techologique vol. 4 1823–35
Edinburgh Encyclopaedia 1830
Encyclopaedia of the Arts 1966
Encyclopaedia Britannica 1794, 1910, 1973
Encyclopaedia of Dates and Events 1974
Encyclopaedia Londoniensis 1810
Encyclopaedia of Photography 1879
Encyclopaedic Dictionary 1882
English Encyclopaedia 1854
Everyman Dictionary of Pictorial Art 1962
Everyman Encyclopaedia 1967
Focal Encyclopaedia of Photography 1969
Harmsworth Encyclopaedia 8 vols
Harmsworth Universal Encyclopaedia 1922
Hutton's Mathematical Dictionary 1796
Imperial Dictionary of the English Language 1898 (ed. J Ogilvie)
Jamieson's Dictionary of Mechanical Science 1827
Knight's Practical Dictionary of Mechanics 1877
McGraw Hill Dictionary of Art 1969
Metropolitania 1845
Oxford Companion to Art 1970
Oxford English Dictionary 1933
Pall Mall of Art 1971
Pantalogia 1813
Penny Cyclopaedia 1836
Physikalisches Lexikon (Leipzig) 1850
Rees's Cyclopaedia 1819
World Book Encyclopaedia 1972
Epstean E trans. (*see* Potonniée G)
Euclid (*see* Danti)

Evening Standard 1979 Fantasy! Dickens at the Aggie, 9 March

*Fabricius, Johann 1611 *De Maculis Sole Observatis* (Wittenburg)
*Farrington, Joseph *Diary* 1793–96 eds K Garlick and A Macintyr (1979)(Yale)
Faulkner T 1829 *Historical and Topographical Description of Chelsea*
Ferguson (*see* Brewster D)
Fink, Daniel A II 1968 An attempt to determine a basis of J Vermeer' method of painting by a comparison . . . *PhD Thesis* Ohio Stat University
*Fladung J A F 1831 *Populaire Vortrage uber Physik Gehalten vor einen Kreise Gebildeter Damen in den Garten von Korompa*
*Fletcher J S 1892–6 The Camera obscura—a story *Chambers Journa* vol. 70
*Foster 1878 *Physics*
*Fouque V 1867 *La Vérité sur l'Invention de la Photographie* (Paris)
Freind, Dr John 1726 *History of Physick* (London)
Fyffe G Bibliography on the camera obscura, includes many Britisl patents, foreign journals and some references to the camera lucid: (Science Library, London)

Gage S H and Gage H P 1914 *Optic Projection* (New York)
Garlick K and Macintyre A (*see* Farrington)
Garnett A E 1911 *The Advance of Photography* (London: Kegan Paul Trench Trubner)
*Gassendi, Pierre 1647 *Instituti Astronomica* (Paris)
Gentleman's Magazine April 1753
Gernsheim H 1969 *History of Photography* (London: Oxford Universit Press)
Gernsheim H and Gernsheim A 1956 *L J M Daguerre: A History of th Diorama and Dagnerrotype* (London: Secker and Warburg)
Gill A J 1976 *Camera Obscura* (Royal Photographic Society Historical Group)
Gioseffi D 1959 *Canaletto—e l'Impiego della Camera Ottica* (Trieste)
Gliozzi M 1932 *L'invenzione della Camera Obscura, Archeion* vol 14
*Goethe J W von 1809 *Die Wahlverwandschaften* (Tübingen)
Goldsheider L 1967 *Johannes Vermeer* (London)
Gowing L 1970 *Vermeer* (London: Faber and Faber)
Grant Production Co 1976 *Catalogue*
*s'Gravesande W J 1711 *Usage de la Chambre Obscura . . .* (The Hague)
Greenacre, Francis *Bristol School of Artists* (Bristol Art Gallery)
Greenwich Royal Observatory
Philosophical Transactions 1790 vol. 80
Hammond, John H 1975 *Communication on Charnock's Drawings*
Howse D 1975 *Greenwich Observatory* (London: Taylor and Francis)

Grigson G and Gibbs-Smith C H 1954 *People, Places and Things* vol. 3 (London: Waverley)
*Grimaldi F M 1665 *Physico Methesis de Lumine* (Bologna)
Gross, Harry I *Antique and Classic Cameras* (New York)
Guillaume de Saint Cloud (*see* Littre)
Guillemin A 1872 *Forces of Nature* (London)
Gunther R T 1925 *Historic Instruments for the Advancement of Science* (London: Oxford University Press)
Guyot E G 1770 *Nouvelles Récréations Physiques et Mathématiques* (Paris)

Halco-Sunbury Co Ltd 1976 *Catalogue*
Hammond, John H 1975 Private communication to the Royal Observatory on Charnock's drawings
—— 1976 Return of the camera obscura *British Journal of Photography* April
—— 1977 Camera obscura *British Journal of Photography* July
**Hannoverische Gelehrte Anzeigen* 1753 No. 44
Hardie, Martin 1966 *Water Colour Painting in Britain* (London: Batsford)
Hardie M and Clayton M 1932 Thomas Daniell and William Daniell *Walker's Quarterly*
*Hardy 1901 *Revue d'Artillerie*
Harris, John 1704 *Lexicon Technicum* (London)
Harris, Joseph 1775 *Treatise on Optics* (London)
*Harrison, W Jerome 1887 *History of Photography* (New York)
Hemlow, Joyce 1958 *History of Fanny Burney* (London: Oxford University Press)
Henderson E 1838 *Arithmetical Architecture of the Solar System* (Glasgow)
*Herigone, Pierre 1642 *Supplementum Cursus Mathematici* (Paris)
Hett W S 1936 *Aristotle—Problems* (London)
*Hevelius, Johann 1647 *Selenographia* (Gedani)
Hilles F W 1929 *Letters of Sir Joshua Reynolds* (London: Cambridge University Press)
Hinton, John 1752 *Universal Magazine* May
Hooke, Robert (*see* Waller R and Derham W)
Hooper Dr W 1787 *Rational Recreations* (London)
Hoppen K T 1970 *Common Scientist in the Seventeenth Century* (London: Routledge and Kegan Paul)
Horizon 1974 vol. 16 No. 3
Howse, Derek 1975 *Greenwich Observatory* (London)
Hutton, Charles 1796 *Mathematical Dictionary*
—— 1803 *Recreations in Mathematics and Natural Philosophy* (London)
*Huygens C 1703 *Dioptrica*

Hypher, Philip 1981 Obscure Camera *British Journal of Photography* 27 March

Imison J 1808 *Elements of Science and Art* (London)
Ingalls A G (ed) 1937, 1963 Amateur telescope making advanced *Scientific American*
Ivins W M Jr 1973 *On the Rationalisation of Sight* (New York)

Jersey Marine (*see* Jones, Barbara); *Western Mail* 30 Dec 1968; 3 Jan 1969
*Jombert, C-Antoine 1755 *Méthode pour Apprendre le Dessein* (Paris)
Jones, Barbara 1974 *Follies and Grottoes* (London: Constable)
Jones, William 1813 Critical observations on Dr. Wollaston's improvements of the camera obscura *Philosophical Magazine* vol. 41
Journals, foreign (*see* Fyffe G)

Kelly, Alison 1975 *The Story of Wedgwood* (London: Faber and Faber)
Kelly R 1976 Great Union camera obscura *Manx Life* (Nov/Dec)
*Kepler, Johann 1604 *Ad Vitellionem Paralipomena* (Frankfurt)
—— 1611 *Dioptrice* (Augsburg)
Kilmarnock
Kilmarnock Standard Annual 1957
McKay A *History of Kilmarnock*
Kilmarnock and Riccarton Post Office Directory 1840
King H C 1955 *History of the Telescope* (London)
Kirch G 1710 *De Divisione Disci Solaris in Camera Obscura, Eclipseos Observandae Gratia* (Berlin Msc. 1, 218)
*Kircher, Athanasius 1646 *Ars Magna Lucis et Umbrae* (Rome)
Kirriemuir Local Authority *Notes on the Camera Obscura Fitted in the Sports Pavilion* (pamphlet)
Klingender F D 1947 *Art and the Industrial Revolution* (Cannington)
Kodak 1971 *Photography* (pamphlet)
*Kohlans J 1677 *Neu-erfundene Mathematische und Optische Curiositaten* (Leipzig)
Koningsberger, Hans 1975 *The World of Vermeer* (London: Time–Life)
Konstam, Nigel 1975 Vermeer's method of observation *The Artist* January
Lambert J H 1773 Expose de quelques observations physiques *Nouveaux Mémoires de l'Académie Royale*
Lardner D 1855 *Museum of Science and Art* vol. 8 (London)
Latimer J 1887 *Annals of Bristol in the Nineteenth Century*
Ledermuller M F 1760 *Microscopischer Gemuths und Augenergotzung* (Nuremburg)
Leurechon J 1627 *Récréations Mathématiques* (English edn 1633)
*Levi Ben Gershon 1342 *De Sinibus Chordis et Arcubis* (German trans. M. Curtz 1901)

Leyser, Baron Ernst von 1842 Erlauternde Worte zu der Camera Clara Dioptrica *Annalen der Physik* vol. LVI (Leipzig)
Libri G 1838 *Histoire des Mathématiques en Italie* vol. 4 (Paris; reprinted Bologna 1967)
*Liesegang F Paul 1919a *Optische Rundschau* Nos 31–33
—— 1919b *Die Kamera Obscura bei Porta. Milleilunger zur Geschichte der Medizin und der Naturwissenschaften* vol. 18
Lindberg D C 1967 Alhazen's theory of vision and its reception in the west *Isis* vol. 58
—— 1968 Theory of pinhole images from antiquity to the 13th century *Archive for History of Exact Sciences* vol. 5
—— 1970 Theory of pinhole images from antiquity to the fourteenth century *Archive for History of Exact Sciences* vol. 6
*—— *Opticae Thesaurus Alhazeni Arabis . . .*
Lindley K 1973 *Seaside Architecture* (London)
Links J G 1972 *Townscape Painting* (London: Batsford)
—— 1977 *Canaletto and his Patrons*
*Littre E 1869 *Histoire Littéraire de la France* vol. 25
Liverpool
Henderson E 1838 *Arithmetical Architecture of the Solar System* (Glasgow)
Llandudno
Liverpool Daily Post 24 Aug. 1966
Llandudno Advertiser 28 Aug. 1966
London, Tower of (*see* Quarrell W H)
Lord, Simon (*see* Robertson)

McGrath, Patrick 1975 *The Merchant Venturers of Bristol*
McKay A (nd) *History of Kilmarnock*
Magazine of Science vol. 1 1839, 1840; vols 3 & 4 1842; vol. 6 1845
Malden R H 1943 *Nine Ghosts* (London: Arnold)
Manx Life Nov./Dec. 1976
Manx Sun 28 May, 29 October 1887
Marbach O and Cornelius C S 1850 *Physikalisches Lexikon* vol. 1 (Leipzig)
*Marci J M 1650 *Disertatio in Propositiones Physico-mathematicus de Natura Iridos, R.P. Balthasarius Conradi* (Prague)
Marek J 1964 An observation of the interference of light of higher orders in 1646 and its response *Nature* vol. 201 Jan.
*—— 1968 Kepler's inventions in physical optics *12th Congrès International d'Histoire des Sciences* vol. 3B 1971
Marion F 1968 *Wonders of Optics* (trans. C W Quin) (London: Samson, Low, Son and Marston)
*Martin, Benjamin 1740 *A New and Compendious System of Optics* (London)

*—— 1772 *Young Gentleman and Lady's Philosophy*
Marx K and Engels F 1970 *German Ideology* ed C J Arthur (London: Lawrence and Wishart)
Mason, William 1819 *The English Garden*
*Maurolico, Francisco 1521 *Theoremata de Lumine et Umbrae*
——1543 *Cosmographia* (Venice)
Mechanics Magazine 1845 vol. 43
Meteyard, Eliza 1866 *The Life of Josiah Wedgwood* (London: Hurst and Blackett)
The Microscope 1892 vol. 12
Middleton, Rev. E 1778 *New Complete Dictionary of Arts and Sciences*
*Molyneux W 1692 *Dioptrica Nova* (London)
*Montucla 1758 *Histoire des Mathématiques* vol. 2 (Paris)
Moore, John 1971 *Brensham Village* (London: Collins)
Moore, Patrick (ed) 1963 *Practical Amateur Astronomy* (Guildford: Lutterworth)
Morton H V 1927 *In Search of England* (London: Methuen)
*Muntz E 1898 Leonardo de Vinci et l'invention de la chambre noire *Revue Scientifique Ser.* 4 vol. 10
Murray A 1808 *Account of the Life and Writings of James Bruce* (Edinburgh)

Nabokov V 1969 *Laughter in the Dark* (original title *Kamera Obskura*) (London: Weidenfeld and Nicolson)
National Gallery, Washington *'A Woman Weighing Gold' by Jan Vermeer* (pamphlet)
Needham J 1954 *Science and Civilisation in China* vols 3, 4 (London: Cambridge University Press)
Negretti and Zambra Ltd 1865, 1879, 1886 *Catalogues*
Newhall, Beaumont 1949 *History of Photography* (New York: MOMA)
Niceron, P J-François 1652 *La Perspective Curieuse* (Paris)
Nicolson M and Rouseau G S 1968 *This Long Disease, my Life: Alexander Pope and the Sciences* (Princetown)
Nollet, Abbe 1752 Chambre obscure de nouvelle construction *Académie Royale des Sciences* vol. 6
*—— 1755 *Leçons de Physique Expérimentale* vol. 5 (Paris)

Ollerenshaw R 1977 Camera obscura in medical illustration *British Journal of Photography* 23 Sept
Oppe A P 1947 *Drawings of Paul and Thomas Sandby* (London: Phaidon)
*d'Orleans (*see* Cherubin)
Oughtred W (*see* Leurechon J)
Ozanam J (*see* Hutton C)

Patents (*see* Fyffe G)

*Peckham, John 1504 *Perspectiva* (Leipzig)
*Peele J 1732 *The Art of Drawing and Painting in Watercolours*
Pepys, Samuel 1666 *Diary* February
Petzval J 1859 On the camera obscura *Philosophical Magazine* vol. 17
*Peuerbach G von 1474(?) *Theories Novae Planetarium* (Nuremburg)
Photographic News Sept 1858; Dec 1859; July 1901
Piffard H G 1892 The camera obscura versus the camera lucida *The Microscope* vol. 12
—— 1893 *British Journal of Photography* 18 Aug
Pirenne M H 1970 *Optics, Painting and Photography* (London: Cambridge University Press)
Playfair R 1877 *Travels in the Footsteps of Bruce in Algeria and Tunis* (London)
Pollack, Peter 1969, 1977 *A Picture History of Photography* (New York: Abrams)
Pope, Alexander 1725 Letter to Edward Blount, 2 June 1725, in *The Works of Alexander Pope* Rev. Elwin Whitwell (1871) (London: Murray)
Porta, G Battista della 1558, 1589 *Magiae Naturalis* (Naples) English trans. 1658, reprinted 1957 (New York: Basic Books)
Potonniée G 1936 *History of the Discovery of Photography* trans. E Epstean (New York: Tennant and Ward)
Priestley, Joseph 1772 *History and Present State of Discoveries Relating to Vision, Light and Colours* (London)

Quarrell W H and Mare M 1934 *London in 1710, from the Travels of Zacharias Conrad von Uffenbach* (London: Faber and Faber)
The Queen's Gallery 1980 *Canaletto: Exhibition Catalogue*
Quin C W trans. (*see* Marion F)

Ralph, Elizabeth 1961 *The Downs: Clifton and Durdham Downs*
Rawlinson H 1953 The camera obscura *British Journal of Photography* April
Rawson P 1969 *Appreciation of the Arts No. 3: Drawing* (London: Oxford University Press)
*Reinerus Gemma-Frisius 1545 *De Radio Astronomico et Geometrico* (Antwerp and Louvain)
*Reinhold, Erasmus 1542 *Theoricae Novae Planetarum . . .* (Wittemberg)
Revue d'Artillerie 1901 vol. 57
Revue Maritime et Coloniale 1869 vol. 25
*Riccioli G B 1651 *Almagestum Novum* (Bologna)
Richardson, Samuel 1754 *Sir Charles Grandison*; 1972 edn (London: Oxford University Press)

Richter J P and Richter I A 1970 *The Literary Works of Leonardo da Vinci* (compiled and edited from the original manuscripts) (London: Phaidon)
*Risner, Friedrich 1572 *Alhazen's Optics*
—— 1606 *Opticae* (Kassel)
Robertson, Scott and Lord, Simon (nd) *The Camera Obscura* (Outlook Tower, Edinburgh)
Roget, John L 1891 *History of the Old Water Colour Society* (London: Longman)
*Rohr M von 1925 *Zentral Zeitung fur Optik und Mechanik* vol. 46
Ronchi, Vasco 1970 *The Nature of Light* (London: Heinemann)
Routledge R 1886 *Discoveries and Inventions* (London)
Royal Academy of Arts 1972 English drawings and water colours 1550–1850 in the collection of Mr & Mrs Paul Mellon *Exhibition Catalogue*
Royal Air Force AML 1674, 1925; AMOs A8–A14 (1932); *Air Publications* 242 1917; 356 1918; 961 1924
Royal Australian Air Force 1943 *Publication No.* 296
Ruskin, John *Modern Painters* vol. 1 (London)

*Salmon W 1671 *Polygraphice* (London)
*Sanderson W 1658 *Graphice* (London)
Scharf A 1968 *Art and photography* (Harmondsworth: Pelican)
—— 1975 *Pioneers of Photography* (London: BBC Publications)
Scheiner C 1619 *Oculus, hoc est Fundamentum Opticum*
—— 1630 *Rosa Ursina Sive Sol* (Rome: Bracciano)
*Schott K 1657 *Magia Universalis Naturae et Artis* (Würzburg: Herbipoli)
Schwarz H 1966 Vermeer and the camera obscura *Pantheon* May/June vol. 24 (an 18th century English poem on the camera obscura; *see* Coke, Van Deren)
*Schwenter D 1636 *Deliciae Physico Mathematiciae* (Nürenberg)
Science and Industry, Annual Record of 1875
Science Museum 1971 *Scientific Trade Cards*
Selwyn E W H 1950 The pinhole camera *Photographic Journal* vol. 90B
Seymour C 1964 Dark chamber and light-filled room: Vermeer and the camera obscura *Art Bulletin* vol. 46
Skopec R (nd) *Photographie im Wandel der Zeiten* (Artia)
Smith R C 1975 *Antique Cameras* (Newton Abbot: David and Charles)
Smith, Robert 1738 *Compleat System of Optics* (Cambridge)
*Society for the diffusion of useful knowledge 1832 *Library of Useful Knowledge, Natural Philosophy* II (London)
Spiller J 1859 The eye as a camera obscura *Photographic News* 23 Dec
Spohr L 1878 *Louis Spohr's autobiography* (London)
Stenger Dr E 1958 *March of Photography* (London: Focal Press)
Sterne L *Tristram Shandy* (London: Dent)

*Storer W 1782 *Storer's Syllabus . . .* (London)
—— 1845 *Mechanics Magazine* vol. 43
*Sturm J C 1676 *Collegium Experimentale sive Curiosum* (Nürnberg)
—— 1701 *Mathesis Juvenilis, Tomus Posterior*
*Sturm, Leonhard C 1708 *Kurzer Begriff der Ganzen Mathesis*
Sutton T 1954 *The Daniells, Artists and Travellers* (London)

Taft R 1938 *Photography and the American Scene* (New York: MacMillan)
Talbot W H Fox 1844 *Pencil of Nature* (London)
*Tarde, Jean 1623 *Les Astres de Bourbon* (Paris)
Tate Gallery 1973 *Landscape in Britain, Catalogue*
**The Tatler* 29 December 1709
Taton R 1963 *Ancient and Medieval Science* (London: Thames and Hudson)
Thorndike, Lynn 1958–64 *History of Magic and Experimental Science* 8 vols (New York)
Tierie Dr G 1932 *Cornelis Drebbel* (Amsterdam)
*Traber Z 1675 *Nervus Opticum* (Vienna)

Uffenbach, Z Conrad von (*see* Quarrell W H)

*Vasari, Giorgio 1809 *Vite de'piu Eccellenti Architetti, Pittori e Scultori* vol. V (Milan)
Venturi 1797 *Essai sur les Ouvrages Physico-mathématiques de Leonardo da Vinci* (Paris)
Victoria and Albert Museum 1972 *From Today Painting is Dead* (London: HMSO)
Vintage Cameras Ltd 1975, 1978 *Catalogues*

*Waller R 1705 *Posthumous Works of Robert Hooke* (London)
Wallis, Mieczyslaw 1954 *Canaletto, the Painter of Warsaw* (Cracow)
Walpole, Horace 1926 *Selected Letters* (arr. by W Hadley) (London: Dent)
Waterhouse J 1901 Notes on the early history of the camera obscura *Photographic Journal* (Appendix contains quotations from the original or early editions of the works of Alberti, Bacon, Barbaro, Benedetti, Caesariano, Cardan, Frisius, Kepler, Leonardo, Maurolycus and Porta)
Watkins and Hill 1838 *Catalogue of Optical Mathematical, Philosophical and Chemical Instruments*
Wells H G 1974 *Land of the Ironclads* (short stories) (London: Benn)
Whatton A B 1859 *Memoirs of the Rev. Jeremiah Horrox . . .* (London)
Wheelock A K Jr 1973 Carel Fabricius *Netherlands Yearbook for History of Art* vol. **24** pp63–83

—— 1977 Constantyn Huygens and early attitudes towards the camera obscura *History of Photography* vol. **1** pp93--103

—— 1978 *Perspective, Optics and Delft Artists around 1650* (New York: Garland)

White J 1957 *Birth and Rebirth of Pictorial Space* (London: Faber and Faber)

Whitley W T 1928 *Artists and their Friends in England, 1700–99* (London: Medici)

Whittemore J 1825 *Historical and Topographical Picture of Brighton*

Whitwell, Elwyn (*see* Pope, Alexander)

*Wiedemann E 1910 *Uber die Erfindung der Kamera Obscura Verhandlunger der Deutsches Physikalische Geselleschaft* vol. **12**

Williams-Ellis, Clough 1971 *Architect Errant* (London: Constable)

Willox J 1856 *The Edinburgh Tourist and Itinerary*

Wills G 1969 *English Pottery and Porcelain* (Guiness Signatures)

Wilson G 1855 Researches on colour blindness: On the extent to which the received theory of vision . . . *Transactions of the Royal Society of Edinburgh* vol. **21**

*Wolf, Professor 1707 *Cours d'Optique* (Halle)

Wolf A 1935, 1938 *History of Science, Technology and Philosophy* vols **1** & **2** (London)

Wollaston W H 1812 On a periscopic camera obscura and microscope *Philosophical Transactions of the Royal Society*

*Wotton, Sir H 1651 *Reliquiae Wottonianae* (London)

*Wurschmidt J 1914 *Sitzungsberichte der Physikalische, Medizinischen Societat in Erlangen* vol. **46**

*——1915 *Zeitschrift für Mathematischen und Naturwissenschaftlichen Unterricht* vol. **46**

Wylde J 1861 *The Magic of Science* (London and Glasgow)

Wynter H and Turner A 1975 *Scientific Instruments* (London: Studio Vista)

Zahn, Johann 1685–86 *Oculus Artificialis Telediopricus Sive Telescopium* (Wurzburg)

The author's collection of notes, references, and material on the camera obscura is now at the Science Museum Library, South Kensington, London.

Index